WOODLAND
WISDOM

WOODLAND WISDOM

A miscellany of forest facts, fiction & folklore

NICK PIERCE

CONTENTS

INTRODUCTION

WELCOME TO *Woodland Wisdom: A Miscellany of Forest Facts, Fiction & Folklore*. If you've spent any length of time in forest or woodland, you'll know that these precious spaces contain endless surprises and stories. This humble guide can only reveal a few of their secrets, but hopefully it will instil a greater sense of wonder and respect for our planet's forest ecosystems in every reader. The ultimate aim is that you come away from this book with an increased desire to explore their history, wildlife and landscapes, whether by devouring other books on the subject or venturing on foot into your local woodland.

This introductory guide is divided into five chapters designed to convey some of the wealth of flora and fauna, life-enriching experience and cultural heritage that forests have to offer. Did you know that 80 million years elapsed between the first plants appearing on land and the very earliest forests? Or that a dead tree, also known as a snag, can still play a vital role in a forest ecosystem? Or that in Norse mythology the first man was said to have emerged from a tree? Or that the 'old men of the forest' – the orangutans of Sumatra and Borneo – share 97 per cent of their DNA with humans? These are just some of the fascinating forest facts you'll discover as you go through this book.

The first chapter begins by outlining the essential characteristics of forest and woodland, including how they first evolved in Earth's prehistoric past and the diverse types that can be found across the globe today. There's a breakdown of the different parts that make up a tree's physical structure and a brief spotter's guide to some of the most common and iconic species.

After this, the focus is on the fungi, insects, birds, reptiles and mammals that make their home in forests around the world, including a look at how some of these species help to create the conditions that allow forests to thrive. This includes the mycorrhizal fungi that form a symbiotic relationship with trees, extending their root systems in exchange for food, and the pollinating and waste-eating bugs that enrich the soil and aid in plants' reproductive cycle.

The third and fourth chapters adopt a more hands-on approach, suggesting some creative and practical ways of engaging with forests and their treasures. There are simple projects such as a step-by-step guide to making a bird house, plus introductions to the arts of whittling and flower pressing. For the more adventurous, there's a guide to foraging for common and not so common forest delicacies, including truffles, chanterelle mushrooms or the mysterious wormwood that is a key ingredient in absinthe, plus information about old folk remedies made from these that are still useful today. There's also advice on what to do if you get lost in the forest, such as to beware following any path you find, believing that it will lead you back to civilization, since many of these are actually the result of grazing sheep!

The final chapter embarks on a whirlwind tour of world mythology and our ancestors' enduring fascination with the storytelling potential of forests. Find out how in ancient Greece, various types of trees were believed to host different kinds of tree nymphs, whereas for the Vikings, the universe itself was a giant ash tree. See how William Shakespeare was inspired by the gigantic forest that surrounded his childhood home and appreciate how woodland is central to J. R. R. Tolkien's conception of Middle-earth.

Read on for an exhilarating investigation into the wonders and ancient wisdom of the woods.

CHAPTER ONE

WOODS, TREES & FORESTS

'In all things of nature there is something of the marvellous.'
ARISTOTLE

'The richness I achieve comes from nature, the source of my inspiration.'
CLAUDE MONET

'The fairest thing in nature, a flower, still has its roots in earth and manure.'
D. H. LAWRENCE

'Trees are sanctuaries. Whoever knows how to speak to them, whoever knows how to listen to them, can learn the truth.'
HERMANN HESSE

What is a forest? How does a tree grow? How can you tell a beech from a birch or an ash from a yew? These may appear to be very basic but, when it comes to woodland, there are no foolish questions. Another name for this chapter might be, 'Everything you ever wanted to know about forests, but were too embarrassed to ask'.

This chapter considers the essential characteristics that make a forest a forest and how they first evolved on land. It decribes a few of the prehistoric forests that have been preserved in petrified form to the present day, and outlines the different types of forest that exist across the globe today.

After that, the focus is on a tree's life cycle, looking more closely at its inner structure and workings, from roots to leaves and everything in between. There's then a concise guide to identifying some of the commonest species of tree and a glimpse at the ingenious scientific processes used to reliably estimate the age of a young or ancient specimen.

Finally, there's a brief look at some of the world's most outlandish forests, from the fantastical to the oddly sinister, including giant baobabs, crooked pines and submerged spruces.

So, pull on your hiking boots, pack a thermos and prepare for an eye-opening woodland odyssey.

WOODLAND & FORESTS

In layman's language, the words 'woodland' and 'forest' are often used interchangeably to mean any area with lots of trees. But for geographers they have more specific applications.

Woodland is a forest with an open canopy. This means that the tops, or crowns, of trees, which are the highest level of foliage in a forest, allow large amounts of sunlight to penetrate to the forest floor. In a true forest, there are larger expanses of vegetation with denser foliage and closed canopies. True forest will often be surrounded by woodland, and often they'll contain the same trees, only packed together more densely in the true forest.

Woodland is often a transitional zone between different types of ecosystem, including true forest, but also desert and grassland. For example, xeric woodland is woodland that borders a desert and is filled with succulent plants, such as cacti, which are able to store water to withstand the dry conditions.

The principal element of both woodland and forest is that trees are the dominant lifeforms in these ecosystems, which are some of the most complex

Woodland

in the world. Both forests and woodlands have something called vertical stratification: this means that there are different layers of vegetation and animal life from the ground layer to the tops of the trees. The nature of these layers differs depending on the amount of light reaching the forest or woodland floor.

FORESTS HAVE:

* A ground layer, with lichens, mosses, fungi and decaying organic matter overlaying the soil.
* A middle layer of plants such as shrubs, which is often spotty and sporadic – or sometimes even entirely absent.
* An upper layer of trees rising up, often tens of metres, and ending in the canopy layer.

ANIMAL LIFE

The diverse species that form part of woodland and forest ecosystems tend to be adapted for vertical movement up and down the different layers in order to seek out the shelter and food necessary to survive. Given the lack of light, particularly in dense forest, these animals are also adapted to compensate for the darkness with an acute sense of hearing, so that they're not surprised by predators and are able to navigate by sound when hunting prey.

Forest

A WORLD OF FORESTS

HOW MUCH OF THE EARTH IS FOREST?

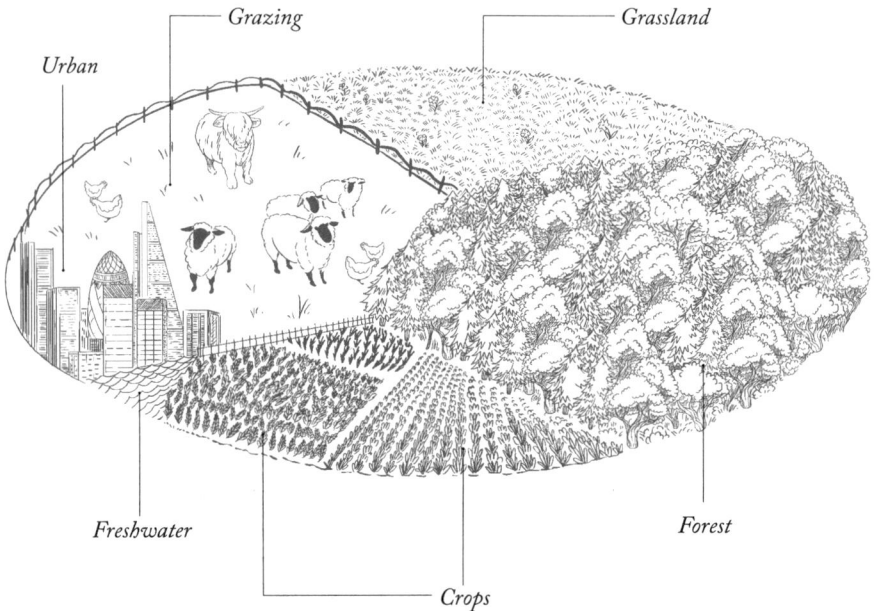

* Forests cover roughly one-third of the Earth's land surface.
* Estimates by experts differ, but forest probably covers around 11½ to 29 million square miles (30 to 75 million km sq).
* Forests are home to over 80 per cent of the world's land-based animals, plants and insects.

FOR A FOREST TO GROW:

* The temperature has to exceed 10°C (50°F) in the warmest months of the year, as trees cannot grow below this temperature.
* There needs to be more than 8in (20cm) of annual precipitation.

MAIN TYPES OF FOREST

There are many kinds of forest all over the globe. They can vary enormously, but experts divide them into three main types: temperate, tropical and boreal.

TEMPERATE FOREST

This is found in eastern parts of North America and across Eurasia. At these latitudes, the year has four distinct seasons (you're probably already familiar with spring, summer, autumn and winter), so the temperature of these forests can alter dramatically, with cold winters and hot summers.

The yearly average temperature is 10°C (50°F) and daily temperatures can range between -30°C (-22°F) and 30°C (86°F).

There's a lot of rain, which means the nutrient-rich soil is able to support a variety of tree types such as birch, oak and maple, as well as mosses, shrubs and herbs. The trees have evolved thick bark to protect themselves during the winter, and they often go into a period of dormancy – or sleep – at the same time.

Temperate forest floor

TROPICAL FOREST

This is found near the equator, in Central America, sub-Saharan Africa and Southeast Asia. Temperatures range between 20°C (68°F) and 31°C (88°F).

These forests are famous for their enormous biodiversity, including numerous rare and endangered species of flora and fauna. Tropical rainforests are so named because of their heavy year-round rainfall, which sustains the broad-leaved evergreen trees that dominate.

There are also tropical dry forests – otherwise known as monsoon forests – where a long dry season of six or more months of the year alternates with a rainy season. These are characterized by deciduous trees that shed their leaves during the dry season and grow new ones at the start of the rainy season.

Another type of tropical forest is the mangrove forest: an ecosystem of trees and shrubs growing in salty or brackish water.

Mangrove trees

{ 16 }

Coniferous forest

BOREAL (OR TAIGA) FOREST

These massive forests can be found across Alaska and Canada, Scandinavia and Siberia in the Arctic region. Temperatures are, on average, below freezing. Typical tree species include pine, conifer, fir and spruce.

Boreal forests play a large role in removing carbon dioxide (CO_2) from the planet's atmosphere.

TYPES OF BOREAL FOREST INCLUDE:

* Coniferous forest: dominated by evergreen trees that keep their leaves all year round, such as pines and firs. These are often uniform in height. Such forests are often located in mountainous regions.
* Deciduous forest: dominated by deciduous trees that shed all their leaves during one season, such as chestnuts, aspens, beeches, birches, maples, elms and oaks. These trees vary in height and shape, resulting in dense growth and less uniformity than can be found in a coniferous forest.

THE EVOLUTION OF FORESTS

Plant species first evolved to live on land roughly 470 million years ago, but it took another 80 million years for forests to emerge. In the meantime, those prehistoric plant species developed the traits that would lead to the evolution of trees, which then outcompeted other species.

PUTTING DOWN ROOTS The first terrestrial plants evolved from species of algae that had previously floated in the planet's early oceans. To overcome the problem of drying out on land, these first plants developed:

* Roots that would fix them to the ground and let them suck up water.
* Tissues that allowed this water to be transported throughout the plant's structure.
* Thick and waxy outer layers that would prevent water loss.

Cooksonia

EARLIEST KNOWN FOREST

Discovered in Cairo, New York this ancient forest is 385 million years old. Experts studying the fossilized remains have uncovered evidence of roots, leaves and wood – all features of trees today.

ARCHAEOPTERIS (385 MILLION YEARS OLD) This prehistoric plant in the Cairo fossil forest resembled modern trees, including woody roots and branches with leaves.

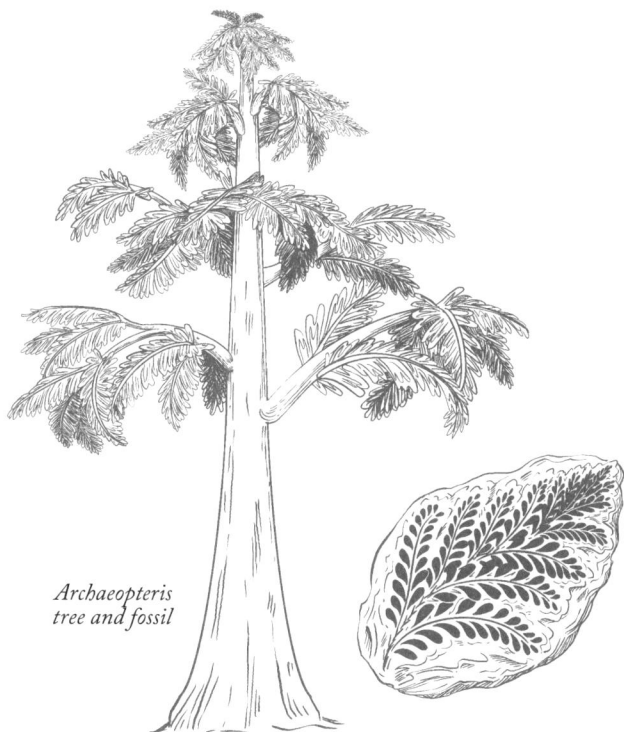

Archaeopteris tree and fossil

BETTER TO BE A TREE It turns out that trees' characteristics gave them an evolutionary advantage in the prehistoric world. The complex branching that trees are capable of meant that they could rise in height above other plants, soaking up more of that precious sunlight for photosynthesis.

IT'S A GAS Changes in the composition of Earth's atmosphere also made a difference to trees' fortunes. At the end of the Devonian period, about 360 million years ago, the level of carbon dioxide (CO_2) in the air began to drop. Prior to this, there was so much CO_2 around that trees couldn't grow the megaphylls – large leaves with branching veins – that are common today. All that CO_2 caused stifling temperatures, so megaphylls would have absorbed too much sunlight and overheated.

Once the CO_2 began to diminish, temperatures dropped and those big leaves became an advantage instead of a liability. Now the trees could suck in CO_2 for use in photosynthesis.

THE CARBONIFEROUS PERIOD (360–300 MILLION YEARS AGO)

Early forests dominated during this time, although their various species would look very alien to our eyes.

LYCOPSIDS These extinct plants, also known as the 'giant club mosses', were truly gargantuan. *Lepidodendron*, one such species, grew up to 115ft (35m) tall and was anchored in place by branching rootlets.

SPHENOPSIDS Living beneath the lycopsids (and also long extinct), these thrived in the damp and swampy landscape of the Carboniferous period. They looked rather like modern horsetails, which evolved from them, and reproduced by releasing spores, rather than seeds.

The first seed plants could also be found in these sweltering forests, including *Cordaites*, with its 98ft (30m) height and upwards-growing branches.

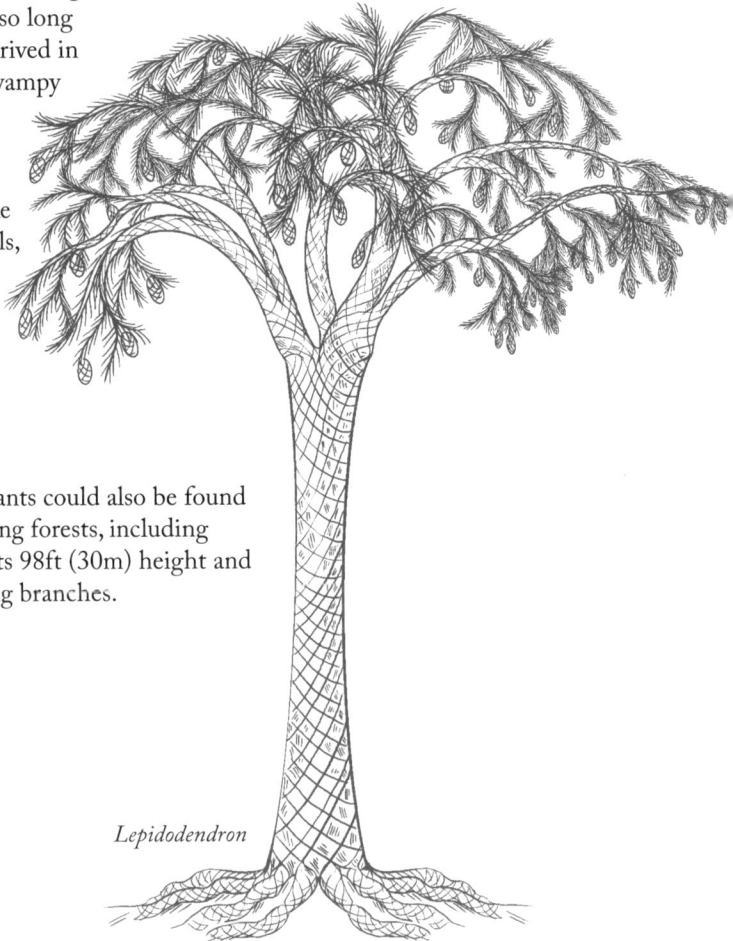

Lepidodendron

A SECOND LIFE

We owe the modern world to the existence of these forests. They eventually collapsed as the climate grew cooler and drier, were buried under soil and then compressed and heated into coal over millions of years. Thanks to this abundant fossil fuel, the Industrial Revolution got underway in the 18th and 19th centuries, transforming every aspect of human life. Ironically, it's the use of the remains of these ancient forests that is one of the factors driving the world towards a climate emergency and threatening the existence of forests today, though many countries are now adopting greener forms of energy.

COAL POWER

* **1769** First coal-powered steam engine patented by engineer James Watt.

* **1882** World's first coal-fired power station opens at Holborn Viaduct in London.

* **1960** Percentage of country's electricity generated from coal power: UK 90%, USA 53%.

* **2023** Percentage of country's electricity generated from coal power: UK 2%, USA 17%.

FROZEN FORESTS

It's not just the bones of dinosaurs and other ancient creatures that have been preserved for millions of years under soil, sea and rock. Entire forests have been fossilized, too.

PETRIFIED FOREST NATIONAL PARK, USA

Petrified Forest National Park in Arizona is one of the world's best places to see remnants of the ancient trees that once covered large regions of Earth's surface. During the Triassic period, 225 million years ago, Arizona was home to lush, tropical rainforests, bursting with plants called ferns, horsetails and cycads, as well as towering conifer trees.

As the land and climate changed over millions of years, sculpted by water and wind, these forests gave way to the barren, rocky desert you'll find here today. But look closely while wandering through this national park and you'll discover that many of the glistening rocks are actually tree trunks, branches and leaves. These Triassic fossilized – or 'petrified' – trees have been buried under layers of sediment from lakes and rivers, absorbing water and silica (a natural substance found in rocks, clay, sand and volcanic ash), which crystallized over time into a hard, glittering mineral known as quartz. Around 40 different minerals have been found in the petrified wood discovered around the world, including silica, calcite and pyrite.

GILBOA FOREST, USA

The oldest known fossil forest dates back to the Devonian period, 385 million years ago – it was discovered in the Catskill Mountains in New York State in the 1920s. Fossils of hundreds of large tree stumps were uncovered in a quarry as workers extracted rock to build the nearby Gilboa Dam. The incredible Gilboa Forest revealed fossils of some of Earth's oldest trees, along with animals that lived in the ancient sea bordering the forest.

SONOMA COUNTY, USA

In the petrified forest in Sonoma County in California, you can see petrified trees from the Pliocene that are 3.4 million years old. When nearby volcano Mount St Helena erupted, the powerful blast knocked down the giant redwoods, burying them in silica-rich volcanic ash. Water filtered through the ash, saturating the trees with silica, which perfectly preserved the wood and created a three-dimensional fossil in a process known as 'permineralization'.

BORTH, WALES

In Wales, the petrified remains of an ancient forest can be seen stretching across the beach at Borth, where the stumps of pine, alder, oak and birch trees rise up from the sand. Believed to be up to 5,000 years old, this submerged woodland remained preserved in peat (decayed vegetation) until a massive storm blew away tons of sand and revealed the fossilized wonder in 2014.

Petrified Forest National Park

LIFE CYCLE OF A TREE

Over millions of years, trees have evolved a system of reproduction that relies on seed dispersal. It's an ingenious system: these hardy little delivery vehicles can be distributed by the wind, water or in animal fur and faeces, ensuring that the next generation of trees can spread far and wide.

GERMINATION OF AN ACORN

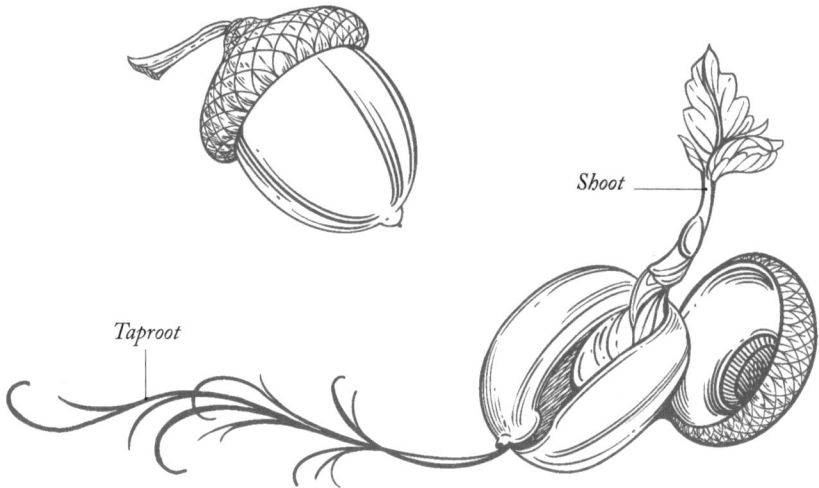

Shoot

Taproot

SPROUT (GERMINATION)

Once a seed has hitched a lift – using the weather or wildlife – and found its new home, it anchors itself in the soil by growing a first shoot, the taproot, which it also uses to draw in water.

Depending on the species, a seed might need certain conditions to be met before it can sprout, such as cold and moist weather or scarification – the breaking down of the seed's hard outer shell so that the shoot can emerge.

SEEDLING

The seed then puts out its embryonic shoot. Once this shoot breaks through the soil and is above ground, it has become a seedling.

This is a vulnerable stage in the new tree's life, when it's at risk of serious damage or death caused by diseases, inclement weather or grazing animals. Ideally, seedlings will emerge near natural items like logs that can shelter them from wind and rain and increase the moisture of the surrounding soil.

SAPLING

Once this infant tree reaches more than 3ft (0.9m) in height, it's classed as a sapling. Unlike fully grown trees, saplings have smoother bark and bendy trunks, and they cannot put forth fruit or flowers.

Some species spend longer than others at the sapling stage. Long-lived trees such as yews and oaks stay saplings for a greater period of time.

MATURE

Once a tree has reached an age where it can start producing fruit and flowers, it is classed as mature. Once these fruits, which contain the tree's seeds, begin to be dispersed, the reproduction cycle can begin all over again. But that's not the end of the cycle for an individual tree.

It takes roughly 12–15 years for the Douglas fir to enter its mature stage, but when it does it can produce seeds for up to 1,000 years.

ANCIENT

Once a tree passes through maturity and is older than most other trees of its species it has become ancient. Depending on the species, a tree might reach this stage in a few hundred years or a few thousand. Identifying traits of an ancient tree include a wide, hollow trunk and a small canopy.

SNAG

When a tree is dying or dead, we call it a snag. Although such a tree has reached the end of its usefulness as a progenitor of the species, it still plays a vital role in the woodland's ecosystem and biodiversity. The dead wood of a snag provides an excellent home for various insect and fungus species – and these creatures are an invaluable food source for larger predators. Similarly, as holes and hollows open in the dead wood, small birds and mammals can use the tree as a shelter from the elements.

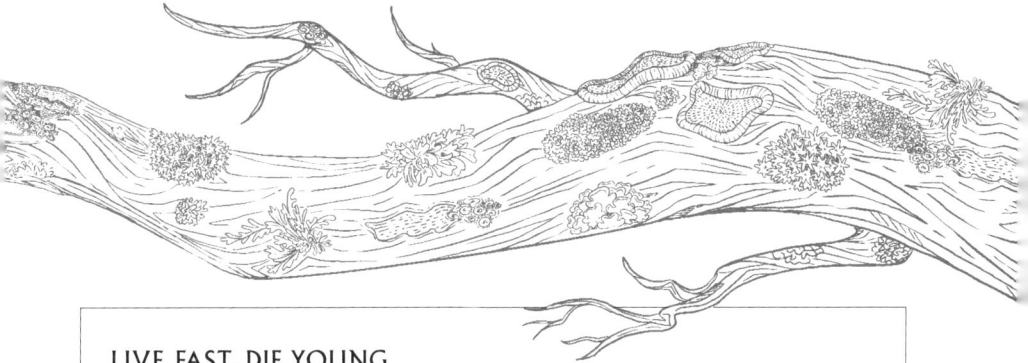

LIVE FAST, DIE YOUNG

Generally speaking, the faster a tree grows and matures, the shorter its lifespan. And this poses problems for the theory that the planting of trees can be used to remove carbon dioxide (CO_2) from the atmosphere and counteract climate change. Although tree growth has been accelerated by the large amounts of CO_2 in the air, especially in urban centres, tree planting tends to be of species that will only last decades or centuries – when they die and decay, all that CO_2 will be released back out into the atmosphere. The answer then, besides emitting much less CO_2, is to ensure that long-lasting trees are being planted, so that carbon sequestration can be guaranteed in the long term.

KNOW YOUR ROOTS

They're not the most aesthetically pleasing part of a tree, but the roots are absolutely essential to its survival. Without these knotty, subterranean organs, a tree could not:

* Anchor itself in the ground to withstand bad weather, such as high winds or flooding.
* Absorb water and minerals from the soil and conduct them to the tree's stem.
* Store reserves of food for times of need.

Tree roots can also produce defensive compounds to help a tree fight back against pests, such as wood-eating insects, and hormones to control when a tree sheds its leaves.

THERE ARE TWO MAIN TYPES OF TREE ROOTS:

ANCHOR ROOTS

Anchor roots fix a tree in the ground to stop it from falling over. These grow much deeper than the other type of root (see below). Once they reach a certain distance from the tree, they can put out additional 'sinker' roots that grow straight downwards, providing extra strength. Anchor roots can last for many years.

FEEDER ROOTS

Feeder roots absorb oxygen, water and minerals into the tree. They can stretch out for metres and metres around a tree to have as wide a surface area as possible, so that there's a greater potential availability of essential nutrients. They grow in the upper 2–2¾in (5–7cm) of the soil where there's access to air and moisture, which are unlikely to be present deeper down.

These feeder roots work on the basis of osmosis: because they contain slightly more salt than the surrounding soil, water will flow into them because nature always seeks a balance.

From there, the tissues in the tree carry these nutrients to where they're needed. Xylem tissues carry water to the leaves, while phloem tissues carry sugar and other nutrients.

ROOT RECORDS

✹ **DEEPEST:** A wild fig tree at Echo Caves in Mpumalanga, South Africa is reported to have roots stretching down 393ft (120m).

✹ **LARGEST:** A grove of quaking aspens in Utah is the largest organism, judged by mass, on Earth. This is due to its enormous root system: 50,000 genetically-identical stems protrude from this same root system, stretching over 100 acres (40.5 ha) and weighing 6,000 tons.

Just like the Ents in J. R. R. Tolkien's *Lord of the Rings,* there really are trees that can walk — although maybe not quite so fast as the ones in that fantasy epic! The palm trees in the Sumaco Biosphere Reserve in Ecuador have developed this ability to deal with soil erosion. They put out new, long roots — sometimes up to 66ft (20m) long — to find firmer ground. As these roots settle in the new spot, the tree bends towards them and the old roots lift slowly into the air. Root growth can reach speeds of ¾–1¼in (2–3cm) a day and it takes a couple of years for the whole tree to relocate to a better location.

INSIDE THE TRUNK

BARK

A tree's bark is a protective layer of organic armour against the world outside, insulating the tree's trunk and branches against cold or hot temperatures, keeping out moisture when it rains and trapping moisture inside when the weather is excessively dry. It also defends the tree against attacks by insects.

Beneath the outer bark, there's a layer of inner bark, also known as phloem. This transports nutrients around the tree. The inner bark only lives a short while, soon hardening and joining the outer bark, while a new layer of phloem replaces it.

CAMBIUM CELL LAYER

This is the wood- and bark-producing factory layer of the tree trunk. Without it, the tree couldn't grow or renew its bark. Hormones transported from the leaves to this cell layer via the phloem stimulate the growth of new cells, which develop into wood and bark.

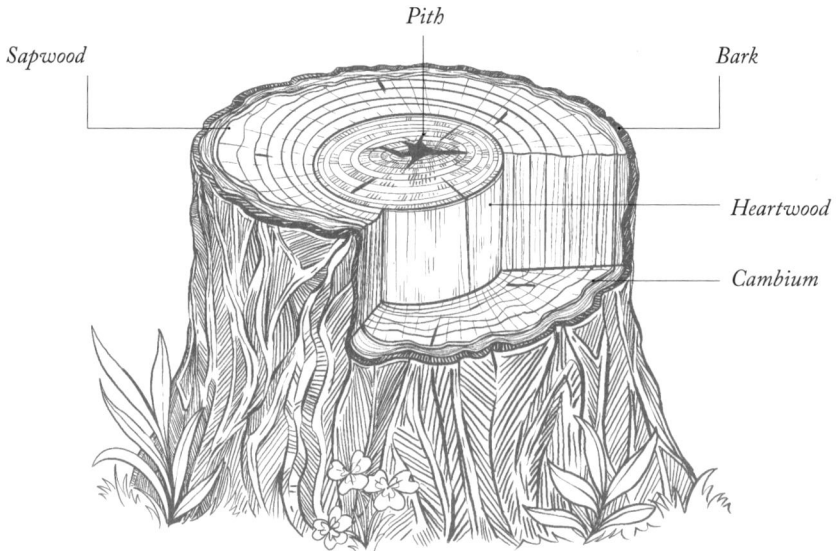

Pith

Sapwood

Bark

Heartwood

Cambium

SAPWOOD

The new wood produced by the cambium cell layer is called sapwood, or xylem. This is the pipeline through which sap – a liquid composed mainly of water, along with sugars and minerals – travels up from the roots to a tree's leaves.

HEARTWOOD

At the centre of the tree is this dead wood – it's what becomes of sapwood when its cells lose their vitality. Heartwood acts as the tree's spinal column, supporting its weight and keeping it upright. The fibres in the heartwood, bound together by a glue called lignin, are strong as steel.

PITH

During the sapling stage, the pulpy pith at the heart of the tree provides nutrients. As the tree ages, the pith darkens and dries out.

BRANCHING OUT

Unlike the roots, it's impossible to miss a tree's trunk and branches. But beneath that innocuous bark there's a highly sophisticated system keeping a tree healthy.

The trunk and branches have two purposes: to elevate the leaves as high above the ground as possible to allow access to sunlight needed for photosynthesis, and to facilitate the transport of food and water around the tree.

ABOVE-GROUND STRUCTURE OF A TREE

BOLE: The part of the tree from the base of the trunk to the first branch.
TRUNK: The central structure of a tree from which the branches grow.
BRANCHES: Typically growing horizontally outwards or upwards from the trunk, the tree's foliage grows from these. Large branches are known as boughs and short ones as twigs.
FOLIAGE: The term for the tree's collection of leaves, needles or scales.
CROWN: The topmost part of the tree.

BIG TREES

GENERAL SHERMAN, USA

The largest living tree in the world by volume, General Sherman looms over Sequoia National Park in the Sierra Nevada mountains of California, USA. It stands 274ft 3⅓in (83.6m) tall, with a diameter of 27ft (8.25m), and is estimated to be around 2,000 years old. It was named after the American Civil War general William Tecumseh Sherman in 1879, at which time it was already considered among the largest trees in the world.

HYPERION, USA

Although its volume makes Sherman the largest tree, the tallest tree overall is also located in California: a coastal redwood named Hyperion, in the Redwood National Park, soaring 380ft 9¾in (116.07m) when last measured in 2019. If you're planning on taking a trip to check out Hyperion's height for yourself – don't bother! The exact location of the tree has been kept secret to prevent tourist footfall from damaging the ecosystem around it, and from 2022 it became illegal to get too close to this formidable redwood.

SAGOLE BAOBAB, SOUTH AFRICA

When it comes to girth, this South African baobab is unbeatable. It has a trunk diameter of 35½ft (10.8m) and a circumference of 108ft (32.89m). Unlike Hyperion, visitors are welcomed, although there's an entry fee to view it up close.

LINDSEY CREEK TREE, USA

The largest tree ever recorded by humans, this coastal redwood in Fieldbrook, California was 390ft (118.87m) tall and had a volume of 90,000 cubic feet (2,550 cubic metres). Sadly, it was felled by a storm in 1905 and no photographs or traces of it survive, although it lives on in legend!

DOUGLAS FIR, SCOTLAND

The tallest tree in the UK is a non-native Douglas fir in Reelig Glen in the Scottish Highlands, which was planted in the 19th century. Part of a stand of huge Douglas fir trees, the tallest has reached a height of 218ft (66.3m).

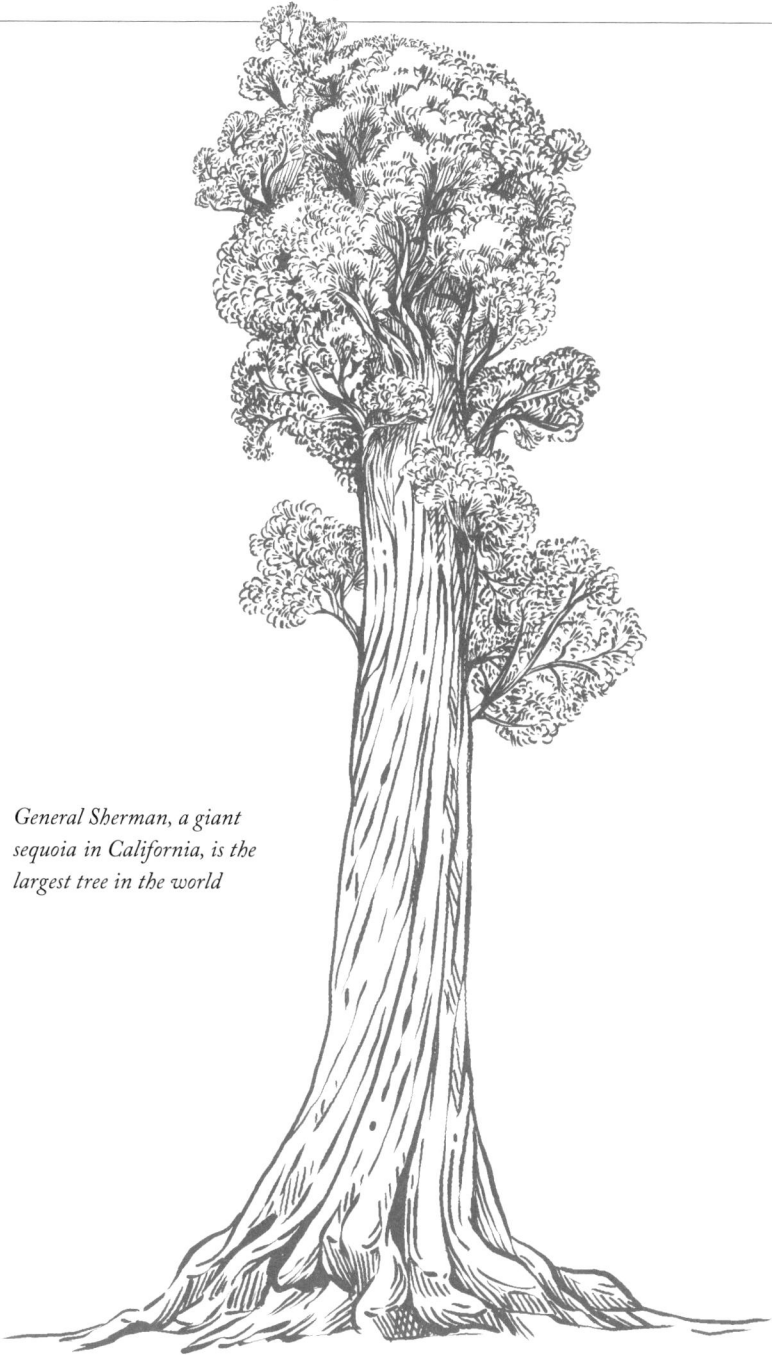

General Sherman, a giant sequoia in California, is the largest tree in the world

GREEN ENERGY

A tree's leaves are integral to the process of photosynthesis, without which it couldn't survive – and without which the planet's atmospheric levels of oxygen and carbon dioxide would be substantially different.

It's important to bear in mind that the needles or scales of some tree species, such as the conifers, are also considered leaves and fulfil the same function.

PARTS OF A LEAF

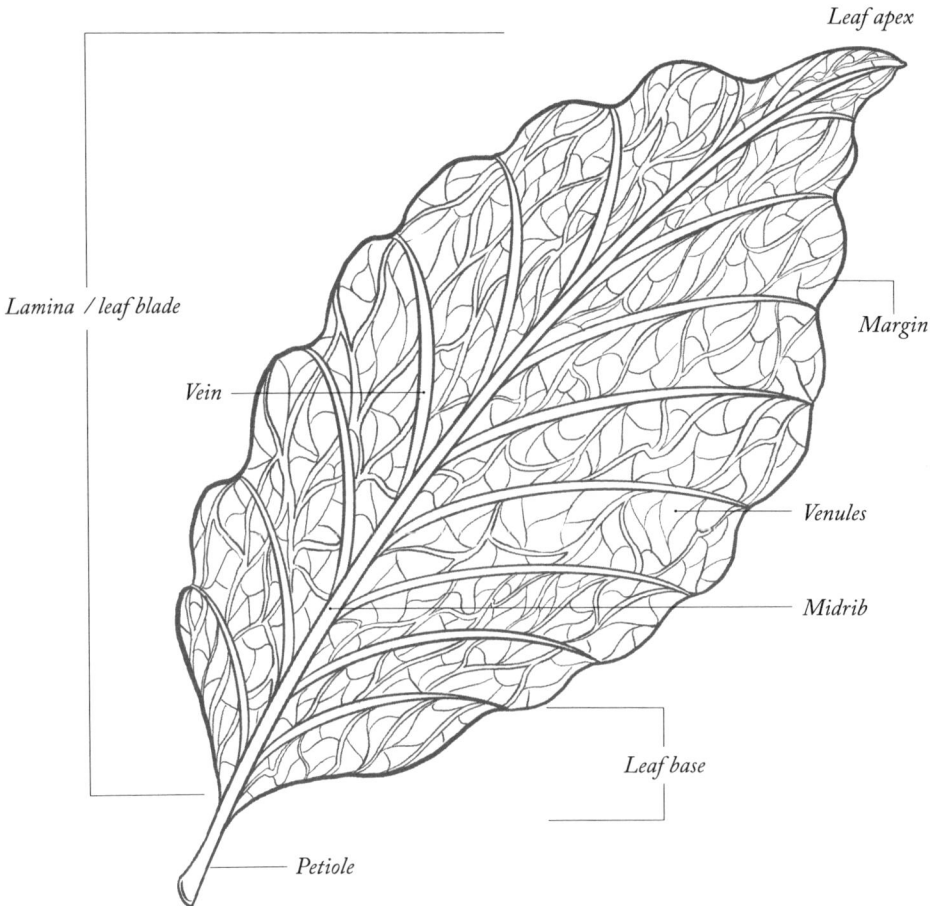

Leaf apex

Lamina / leaf blade

Margin

Vein

Venules

Midrib

Leaf base

Petiole

NATURAL CHEMISTRY

The green pigment in tree leaves – chlorophyll – aids photosynthesis. Chlorophyll absorbs the sun's light so that the leaf can convert carbon dioxide from the atmosphere and water taken up from the soil by the roots into sugar and oxygen. This sugar is used to fuel the tree's growth or stored in its branches, trunk and roots for use at a later date. It's transferred to the tree by the stems that connect each leaf to a twig or branch. Meanwhile, the oxygen is released back into the atmosphere, raising Earth's oxygen levels.

By removing carbon dioxide and other pollutants, trees play a globally significant role in improving air quality and counteracting the causes of global warming.

Although a tree may last centuries, no single leaf lasts very long. They're short-lived structures that are shed and replaced by the tree on a regular basis.

In deciduous trees – such as maples, elms and oaks – the leaves are shed every autumn and new ones sprout the following spring. Evergreen trees – such as conifers including pine and spruce – keep their green leaves all year round. However, this doesn't mean that the leaves are never shed. Unlike in deciduous trees, evergreens continuously shed their leaves and replace them with new ones. It's a less visible and spectacular process than in deciduous trees, but it's no less dynamic.

SURVIVAL PLANS

Deciduous and evergreen trees take different approaches to withstanding changes in temperature and rainfall during autumn and winter.

Deciduous trees tend to have broader leaves, enabling them to manufacture as much food as possible during the summer months so that they have enough energy to last during the colder part of the year. In winter, by shedding their leaves, deciduous trees can shut down processes like photosynthesis that would use up energy that needs to be conserved until more amenable weather returns.

By contrast, evergreens have high nutrient needs all year round. This is so that they can stay in good condition and prevent damage to their essential parts during the poor weather conditions of winter. They tend to have smaller leaves, with a waxy coating that protects them from freezing.

HOW TO IDENTIFY TREES

LEAF SHAPE CAN HELP IDENTIFY A TREE

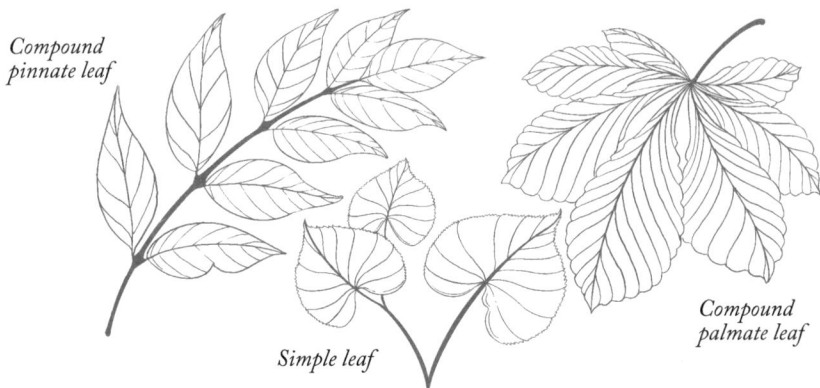

Compound pinnate leaf

Simple leaf

Compound palmate leaf

There are many features of a particular tree that can be used to identify it. As a rule, the more features you can see and take into account, the easier and more accurate your identification is likely to be. Some features to look out for are:

❋ Does it have leaves or needles? If it has needles or scales, it's probably a type of conifer, such as a fir, pine, larch or spruce.

❋ What sort of landscape is it in, e.g. in a field, close to water or in woodland?

❋ What time of year is it? This will affect the tree's appearance and what identifying features you can consider.

❋ Size and shape, e.g. does it have a narrow crown? Does it have a single trunk or many stems? What are the branches like?

❋ What is the bark like? Is it smooth or rough? What colour is it?

❋ What do the flowers look like? Do they contain both male and female parts, or are the male and female reproductive organs on separate flowers – or even separate trees? And what time of year are they blooming?

❋ What is the texture and colour of the tree's fruit? Do the fruits appear singly or in clusters?

❋ In winter, you can examine the leaf buds and twigs. How are the leaf buds arranged on the twig and what is their shape? What is the texture of the twigs, e.g. smooth? Hairy?

ASH (OR EUROPEAN ASH)
Fraxinus excelsior

Native to Europe, Africa and Asia Minor, the ash is the third most common tree in Britain.

IDENTIFYING CHARACTERISTICS:

* It's tall – up to 115ft (35m) in height.
* Bark is pale brown to grey in colour and develops fissures as it ages.
* It has light green, oval leaves.
* In winter, the smooth twigs have black, velvety-textured buds arranged opposite each other.
* It's dioecious: male and female flowers usually – but not always – grow on separate trees. All its flowers are purple, growing in spiky clusters at the end of the twigs in spring.
* The winged fruits, known as 'keys', develop in late summer and autumn.

In Norse mythology, the first man on Earth is said to have emerged from an ash tree. It has long been believed to have supernatural properties, with the wood being burned to deter evil spirits.

An ash tree can live for 400 years.

The tough wood, which can withstand shocks without splintering, is commonly used to produce tools and sports handles, including hammers, axes, oars and hockey sticks. It is also a popular wood for making furniture.

EUROPEAN OAK (COMMON OAK, ENGLISH OAK)
Quercus robur

The mighty oak is an essential part of woodland ecosystems, supporting more species than any other tree, including hundreds of insect species, the birds that feed on them, the small mammals that eat its acorns and the fungi that are nourished by the rich leaf mould that accumulates around it.

IDENTIFYING CHARACTERISTICS:

* A large tree, up to 65–130ft (20–40m) in height, with a broad, spreading crown and thick branches.
* Round-lobed leaves with short leaf stalks, around 4in (10cm) long. The leaves burst out in mid-May and grow in bunches.
* The flowers are long, yellow catkins. They release pollen into the air.
* The fruit are acorns – ¾–1in (2–2.5cm) long – which grow on long stalks. They turn from green to brown as they ripen then drop to the ground below.
* In winter, it has clusters of rounded flower buds – each bud has more than three scales.

In England, the oak tree is a symbol of strength. It was long a tradition for couples to get married beneath an old oak tree.

One of the strongest and most long-lasting timbers on Earth, oak has been used for thousands of years to manufacture furniture, flooring, barrels and much else. The tannin found in the bark has been used to tan leather since at least the days of the Roman Empire.

COMMON BEECH
Fagus sylvatica

These enormous, awe-inspiring trees often grow together in beech woods – a habitat for many varieties of butterflies and fungi, as well as rare plant species, including the orchid red helleborine. They have long been associated with femininity and are seen as the queen of British trees (with oak the king).

IDENTIFYING CHARACTERISTICS:

* They're very tall, reaching heights of more than 130ft (40m), with a vast domed crown.
* Smooth, grey bark, often with faint horizontal etchings.
* The leaves are initially lime green with silky hairs, but get darker and lose their hairs as they mature. They're oval in shape, pointed and 1½–3½in (4–9cm) long.
* It's monoecious: the male and female flowers grow on the same tree, becoming visible in April and May. The male flowers are catkins hanging from long stalks, while the female flowers grow in pairs and are surrounded by a cup.
* Once the cup is pollinated, it grows woody in texture and encloses beech nuts.
* In winter, the leaf buds form on short stalks. They are torpedo-shaped and reddish-brown in colour, with a criss-cross pattern.

In Celtic mythology, Fagus is the god of beech trees.

Beech wood can be used as a fuel source or to make cooking utensils and furniture. It was traditionally used to smoke herring.

The tallest native tree in the UK is a beech: located in Hagg Wood in Derbyshire, it stands 147ft (45m) tall.

YEW
Taxus baccata

Symbols of death and immortality, yew trees were often grown in churchyards. Why? One practical reason for this is their toxicity, which stopped people from grazing their cattle on church grounds for fear of poisoning their animals. They're some of the longest-living trees in the world.

IDENTIFYING CHARACTERISTICS:

* Up to 65ft (20m) in height.
* Reddish-brown and purplish peeling bark.
* Its needle-like leaves, which remain all year round, can be seen in two rows along its twigs. The leaves are dark green above and grey-green below.
* It's dioecious: the male and female flowers grow on different trees. They're visible in March and April. The male flowers are globe-like in shape and yellow-white in colour. The female flowers are bud-shaped, scaly and green when young, turning brown and acorn-like as the season wears on.
* Its seeds are enclosed in red, fleshy, open-tipped structures called arils.

Traditionally, yew branches would be carried on Palm Sunday and during funeral ceremonies.

The Fortingall Yew in Perthshire, Scotland is believed to be one of the oldest living things in Europe, with its estimated age somewhere between 3,000 and 9,000 years old!

The yew tree's highly poisonous foliage actually has a medicinal use. The toxic taxane alkaloids have been harvested from the trees and used in anti-cancer drugs.

SYCAMORE
Acer pseudoplatanus

Native to central, eastern and southern Europe, the sycamore is an extremely fertile tree that can spread easily throughout woodland habitats – often at the expense of other native species in those countries, such as the UK, where it has been introduced by humans.

IDENTIFYING CHARACTERISTICS:

* Grows up to 115ft (35m) tall.
* Dark pink-grey bark, which is smooth in young sycamores but cracked and covered in small plates as the trees mature. Its twigs are hairless and pinky-brown in colour.
* It has five-lobed leaves, roughly 2¾–6in (7–16cm) in length.
* Its flowers are green-yellow in colour and hang in spikes.
* Its famous winged fruits, known as samaras, which spiral down to the ground.

The sycamore has long been associated in cultures across Europe with fertility. In ancient Greece, the sycamore was linked with the goddess Hera, the queen of the gods and the goddess of marriage and childbirth.

The strong, pale cream wood of sycamore is ideal for carving into sturdy, practical furniture and kitchen utensils. Although they're common in woodland, sycamores' resistance to pollution also makes them suitable for planting along suburban streets.

ASPEN
Populus tremula

Also known as quaking aspen, because of the beautiful way that its leaves shimmer in the breeze, this tree is native to ancient woodland in cool regions of the northern hemisphere.

IDENTIFYING CHARACTERISTICS:

* Grows up to 82ft (25m) tall.
* It has grey bark, often marked with diamond-shaped pores called lenticels. In older aspens, the trunk can be covered in lichen, making it look black. The twigs are dark-brown, shiny and slender in shape.
* The highest branches of the aspen are often bent over horizontally.
* Its leaves are round with irregular teeth. They change in colour from coppery to green to yellow or red as they age.
* The flowers appear in March and April. The male catkins are brown, roughly 4¾in (12cm) long and turn yellow when ripe and filled with pollen. The female catkins start green and then mature into fluffy fruits. They open up to release their seeds into the wind in summer.
* In winter, aspens can be spotted by their extremely knobbly twigs, covered in flower buds that spiral around.

Crowns made of aspen are often found in ancient burial mounds.

Aspen's wood is quite fire resistant, making it a popular choice for matches and paper manufacturing. Shredded aspen wood is also often used for packing and stuffing.

SILVER BIRCH
Betula pendula

A symbol of purity and renewal in Celtic mythology, silver birch can flourish in a range of temperatures, making it a tree that has colonized woodlands from the Mediterranean to the subarctic region.

IDENTIFYING CHARACTERISTICS:

* A medium-sized deciduous tree, reaching 98ft (30m) in height. It has drooping branches. The canopy of a silver birch is light and open.

* The bark is mostly white, remaining this way all year round, but black and rough at the tree's base. The bark develops diamond-shaped fissures as the tree matures. The tree's twigs have a smooth texture and are covered in small, dark warts.

* It has small, light green, triangular-shaped leaves with a toothed edge that turn yellow when autumn arrives.

* The male and female flowers can be found on the same tree and appear from April to May. The male catkins are yellow-brown and long, hanging down in groups of two to four, while the female catkins are short, bright green and erect. After pollination, the female catkins swell and turn dark crimson, releasing tiny seeds that are dispersed by the wind.

Hole-nesting birds such as woodpeckers often settle in the trunk of the silver birch. The leaves are a popular meal for the caterpillars of many moth species.

Birch wood is durable and heavy, so it's a suitable material for furniture, children's toys and other everyday items.

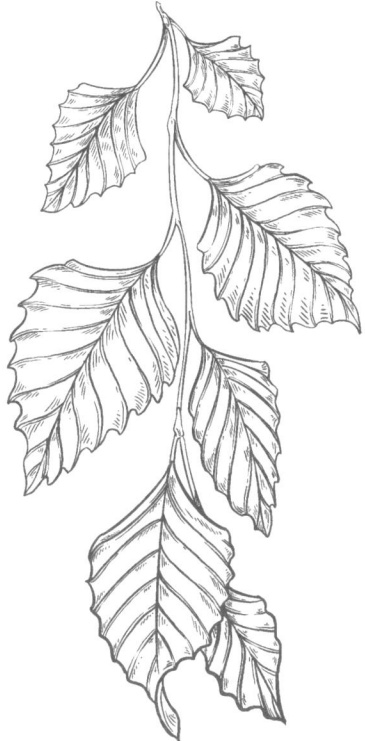

DENDROCHRONOLOGY

Also known as tree-ring dating, this scientific process can be used to accurately measure the age of a tree and garner knowledge about climate and environmental changes over the course of its lifespan.

The science of dendrochronology relies on the fact that many, but not all, tree species produce growth rings inside their cores during annual growing seasons. The wider the ring, the greater the growth in that year, and by analysing the qualities of these rings – and comparing them to ring sequences from other tree cores – a picture of the events that affected tree growth in a particular region over centuries can be assembled. By lining up these ring measurements with those from trees from overlapping ages, experts can construct timelines stretching back for millennia.

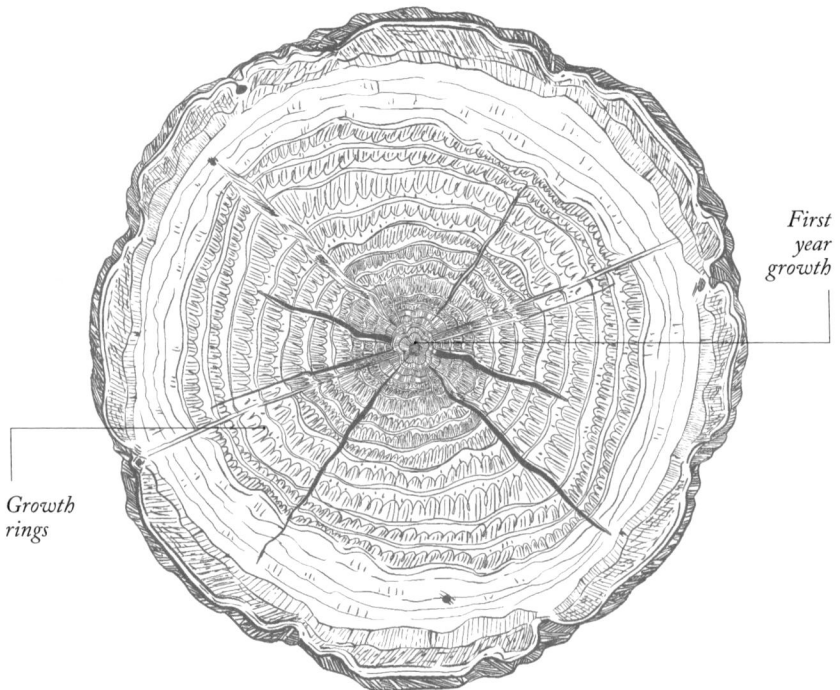

First year growth

Growth rings

Some bristlecone pines in California are more than 4,000 years old, so they're an excellent resource for dendrochronologists to look into the distant past.

RING WIDTH IN ANY GIVEN YEAR IS MOSTLY AFFECTED BY:
* The amount of precipitation
* The prevailing temperatures in the region

Details such as scars, abnormal cells and ring thickness can be used to identify major geological and weather events such as:

* Volcanic eruptions
* Earthquakes
* Forest fires
* Floods
* Landslides
* Avalanches
* Severe frosts

HOW DOES IT WORK?

A device called an increment borer is used. This narrow metallic tube is inserted into a tree to obtain a core sample stretching from the bark to the very centre. This core is later separated out into layers in a laboratory so that the rings can be counted, measured and analysed.

Although dendrochronology is often practised on living or recently felled trees, samples can also be drawn from naturally preserved timbers, such as in peat bogs or river gravels, and from wood used in the construction of old buildings.

A PAINLESS PROCEDURE Don't worry, when carried out properly taking such samples is entirely harmless to the tree. The living tree seals off the small wound from the rest of its tissue ensuring that no lasting damage is done.

WEIRD WOODS

From a forest's worth of preserved dead trees to a magical, mystical grove, these woods from around the world are extra special, drawing visitors to them to experience the weird – and the wonderful.

CROOKED FOREST, POLAND

This grove of 400 peculiar pine trees is a popular tourist attraction. What's so strange about them? Each of the pines is bent at the same strange angle, veering sharply to the north just above ground level for 3–10ft (1–3m) before curving upright. The reason for this odd phenomenon remains mysterious, but it's most likely manmade as it was only planted around 1930. One theory is that the trees were deliberately deformed to supply naturally curved timber for use in the furniture-making and shipbuilding industries.

CLEAR LAKE FOREST, OREGON, USA

Beneath the icy waters of Clear Lake lies the preserved remains of an ancient forest. This strange and stunning natural landscape was formed 3,000 years ago when the nearby volcano Sand Mountain erupted, burning down the forest that grew beside the McKenzie River. As the lava solidified, it dammed up the river, which grew into Clear Lake, submerging the remains of the incinerated forest in its near-freezing depths. You can see the 80ft (24m) trunks of the trees from the surface.

DEADVLEI, NAMIBIA

Like Clear Lake, this is a dead forest, but no less spectacular for it. Nestled in a valley between the sand dunes of the Namib-Naukluft National Park, the trees perished around 600–700 years ago from a lack of water. But the intense desert heat stopped them from decomposing and scorched them black, creating a landscape of great sculptural beauty that is a popular filming location for Hollywood productions.

ARASHIYAMA BAMBOO GROVE, KYOTO, JAPAN

Situated near the Tenryū-ji Buddhist temple, this natural forest of bamboo, covering 6¼sq miles (16sq km), is a much-visited tourist attraction. The rows of soaring bamboo conjure an otherworldly, meditative atmosphere.

GOBLIN FOREST, NEW ZEALAND

Also known as the Kamahi Walk, this strange forest is famous for featuring as a fantasy landscape in Kiwi filmmaker Peter Jackson's *Lord of the Rings* movies. It owes its unearthly magic to its abundance of kamahi trees, which have grown on the trunks of other tree species. Over time, the trunks and branches of the kamahis have entwined with those of the other trees, creating the sort of knotty and twisted forest that is reminiscent of fairy tales. The effect is heightened by the liverworts, ferns and hanging mosses clinging to the trees.

ANCIENT BRISTLECONE PINE FOREST, CALIFORNIA, USA

Sitting high in the White Mountains of Inyo County in eastern California, this protected forest is notable for its bristlecone pines. These remarkable trees grow at high altitudes – 9,800–11,000ft (3,000–3,400m) above sea level – have oddly contorted and knotted bodies and live for an exceptionally long time. As of 2024, the Methuselah pine in one grove here was 4,856 years old, making it the world's oldest known and confirmed non-clonal living organism.

Crooked Forest

HALLERBOS, BELGIUM

It's not the trees that have given Hallerbos its reputation, but the sprawling carpet of bluebells that blooms across the 1,360 acres (552 ha) of forest floor for a few weeks each year. Visitors come from far and wide to witness this explosion of colour.

DRAGON BLOOD TREES, SOCOTRA, YEMEN

Native to the Socotra archipelago in the Arabian sea, dragon bloods are one of the world's weirdest species of tree. For one thing, they have a bizarre shape, their upturned and densely packed crown resembling an umbrella. But that's not all: they also produce a deep red resin that looks like blood. This resin has been used as a traditional medicine for thousands of years for ailments such as fever, ulcers and diarrhoea.

AVENUE OF THE BAOBABS, MADAGASCAR

It's the imposing scale of this grove of Grandidier's baobabs, lining 850ft (260m) of an unsurfaced section of the RN8 highway, that makes it so striking. Each of the baobabs is roughly 98ft (30m) in height. They were once part of

Avenue of the Baobabs

a dense tropical forest, but the rest of the vegetation was cleared to make way for agricultural land, and now these noble giants stand alone.

HOIA FOREST, ROMANIA

Often called 'the most haunted forest in the world', the distinctly spooky Hoia is notorious for its legends of ghosts and other paranormal activity. In recent years, it has been the site of many UFO sightings – and there have been tales of people disappearing, only to reemerge months or years later with no knowledge of what happened to them. The easily frightened might want to give it a wide berth, but for fans of the macabre and supernatural it's surely too good to resist.

SUNKEN FOREST, KAZAKHSTAN

When a landslide caused by an earthquake created a natural dam in 1911, rain and river water slowly formed a new lake and submerged the Asian spruce trees growing here. The limbless, whitened trunks of the trees still stick out of the water, looking like the masts of sunken ships. Beneath the lake, a new ecosystem of algae and other underwater plants has amassed around the trees' lower regions.

CHAPTER TWO

FAUNA & FUNGI

'Never does nature say one thing and wisdom another.'
JUVENAL

'The best time to plant a tree was 20 years ago.
The second-best time is today.'
CHINESE PROVERB

'I love not Man the less, but Nature more.'
LORD BYRON

'Heaven is under our feet as well as over our heads.'
HENRY DAVID THOREAU

The trees are just one layer of complexity in a forest. They interact, in ways big and small, with the other flora and fauna that form a forest ecosystem. These organisms are proof of how life can find ingenious ways to thrive in every available niche, from the fungus that flourishes underground to the birds that roost and sing in the canopy.

This chapter takes a look at some of the species that call forests home, including critters that will be familiar to those who've frequented the woodlands of Europe and North America, and other animals native to more far-flung regions of the world. It also glances at the battle for supremacy between red and grey squirrels, considers why the reintroduction of beavers has been so beneficial for woodland biodiversity and investigates a few weird and wonderful fungi, from the aptly named stinkhorns to the wobbly witches' butter.

And if you've ever wondered what makes the Komodo dragon such a formidable apex predator in the forests of the Lesser Sunda islands, or why the decomposition carried out by insects is so essential to woodlands' wellbeing, or how a woodpecker keeps its brain from turning to mush when it hammers on a tree trunk, just turn the pages to find out. But, be warned: things might get a little wild.

FUNGI

Once classed as plants, the 144,000 known species of fungi are now considered to be their own separate kingdom of organisms. These fascinating and freaky entities thrive in soil and water in mild, moist conditions, which means they're perfectly suited to populating woodland areas.

HOW DO FUNGI WORK?

Most fungi consist of a mass of strands called hyphae. This mass forms the main body of a fungus and is called the mycelium. Some fungi develop an additional part that is used to disperse spores, the cells that allow the fungus to reproduce. This part is the mushroom that we see above ground, while most of the fungus is hidden beneath. Sounds simple enough, but there is huge variation among species in size, shape and appearance.

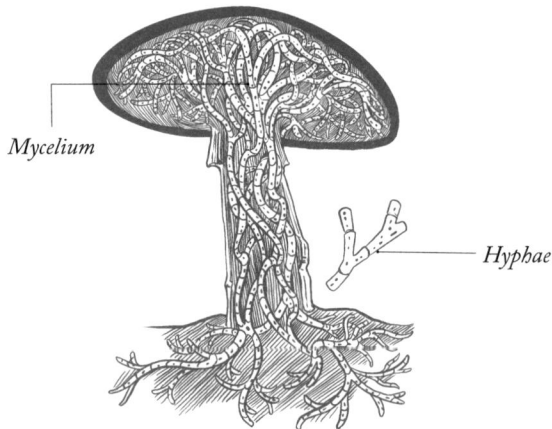

Mycelium

Hyphae

THERE ARE TWO MAIN TYPES OF FUNGI:
Saprophytes, which live by feeding on dead animal and plant matter. They break down this matter and return it to the soil, fulfilling an important role in ecosystems, including woodland ones.

Parasitic fungi, which live off living animals and plants, making those unlucky organisms unwell or even killing them.

MYCORRHIZAL FUNGI

There are types of fungi that cooperate with trees and other plants for the benefit of both organisms. They grow among the plants' roots, taking sugars from them in exchange for the nutrients and moisture that the fungal strands have gathered from the soil. This relationship helps to greatly extend the root system and absorptive potential of a tree or plant.

FLY AGARIC
Amanita muscaria

If you picture a fungus in your mind's eye, it's highly likely that this striking toadstool is what you'll see. With its scarlet or orange cap, stippled with white spots, it looks like something out of a Brothers Grimm fairy tale. It grows in the soil around birch, pine or spruce trees.

Despite its whimsical appearance, fly agaric is no laughing matter. Highly toxic, it was traditionally used as an insecticide. If consumed, it has psychoactive and hallucinogenic effects. It's a likely candidate for the substances used for thousands of years by ancient cultures during shamanic ceremonies to induce a heightened state of consciousness, and it's also thought to be the inspiration for the psychedelic toadstool in Lewis Carroll's *Alice's Adventures in Wonderland*.

Reports of human deaths by fly agaric are rare, but it's best not to mess around with this twee but toxic woodland denizen.

WITCHES' BUTTER
Tremella mesenterica

Wobbly, gelatinous and orange, witches' butter (aka golden jelly fungus/yellow brain/yellow trembler) doesn't resemble your average fungus. Its lobes are greasy to the touch when wet, growing hard when it dries out. Found clinging to fallen branches and dead wood in temperate zones across the UK and Europe, as well as in America, Asia and Australia, it's a parasitic fungus that feeds on other wood-rotting fungus. In folklore, seeing this gruesome-looking fungus on the gate or door of a house means that it has been cursed by a witch and must be pierced with a sharp object to neutralize the spell. Unfortunately once a piece of wood is infected, the fungus will return anyway!

BRACKET FUNGI
POLYPORACEAE

You've almost certainly seen some of these common species, sprouting like shelves from the branches and trunks of both living and dead trees. Some of the more striking specimens include:

Chicken of the woods

Beefsteak fungus, *Fistulina hepatica*, which not only looks rather like a side of sirloin but actually leaks blood-red droplets.

Chicken of the woods, *Laetiporus sulphureus*, which is bright yellow and resembles coral from an ocean reef. It's common on yew and oak trees.

Hoof fungus, *Fomes fomentarius*, which has a dome-shaped upper side and a flat underside. It's greyish in colour and was used as a portable material for lighting fires for thousands of years.

STINKHORNS
PHALLACEAE

The Latin name of the common stinkhorn – *Phallus impudicus* – is highly appropriate, given its marked resemblance to a man's private parts. But that's not where the weirdness ends with this species. After emerging from an egg, the fungus grows and carries a green, spore-carrying coating on its tip called the gleba, which smells like rotten flesh. Flies eat this coating, dispersing the spores and leaving the fungus white.

Reportedly, in the Victorian era, the common stinkhorn's shape so scandalized polite society that gentlemen and older ladies would knock down or remove and burn them to prevent these suggestive fungi from encouraging lewd thoughts in the minds of young women and servants.

INSECTS

From pollination and treating soil to helping dead things break down, insects are an essential part of woodland ecosystems. Subterranean insect activity helps air enter into the soil, which lets stale carbon dioxide out and makes room for fresh new oxygen. This creates fertile soil that drains well and even helps plants thrive and absorb water. Bugs that burrow into soil, such as ants and beetles, help distribute water with the channels they create. Insect excrement, also known as frass, provides nutrients and helps fertilize soil.

FLOWER POWER

Around 75 per cent of all flowering plants need pollinators (animals that help with pollination) to pass pollen from plant to plant – and plants have developed clever ways of luring them in.

Colourful flowers do a great job of attracting pollinators to plants. It's thought bees are more attracted to bright blue and violet flowers, butterflies prefer yellow, orange and pink flowers and beetles like white or cream colours.

Releasing pleasant smells is another way plants draw in pollinators. When it comes to scent, beetles are more attracted to musty or spicy aromas, while bees and butterflies prefer lavender, mint and herby odours.

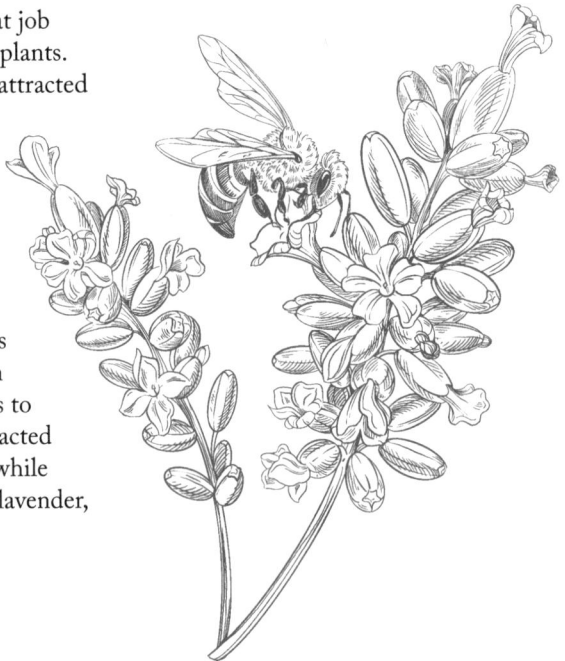

BREAKING IT DOWN

Another important way insects help soil is as decomposers. Also known as detritivores, these scavenger bugs feed on decomposing matter. There are three main groups of detritivore insects: ones that feast on dead or dying plants, others that eat dead animals and those that ingest animal excrement. These hungry scavengers help recycle nutrients back into the soil and create topsoil, the nutrient-rich layer that helps plants grow. These bugs include beetles, blowflies, termites, flies and cockroaches.

MEAT-EATERS

Detritivores are also sometimes known as saprophages. This name comes from the Greek words 'sapros', meaning rotten, and 'phagein', meaning eat.

GIANT MALAYSIAN KATYDID
Arachnacris tenuipes

Found in Malaysia's mountainous forests, these giant crickets stay motionless during the day and use their camouflage to confuse predators. With their green skin and beady black eyes, katydids are one of the largest insects around, with a body of 6in (15cm) long. Their extra-long antennae can even outstretch the length of their bodies, growing up to three times longer.

Male katydids make a high-pitched, shrill sound to attract females by rubbing their wings together, a process called stridulation. They're one of the loudest insects in existence.

UPIS BEETLE
Upis ceramboides

Also known as the Alaskan darkling beetle, this forest-dwelling insect can withstand wintry conditions as cold as -73°C (-100°F). It's able to produce a sugar-based, frost-busting liquid called xylomannan, which lines the membranes of its cells. This is a last resort though – upis beetles will dig deep into aspen and willow trees to avoid the icy weather.

STAG BEETLE
Lucanus cervus

Great Britain's largest beetle, reaching up to 3in (7.5cm) long, stag beetles are easily identified by their red-brown bodies and impressive, antler-like jaws. Found across Europe, but not in Ireland, they live in woodland and are especially fond of oak woods.

Stag beetles spend most of their lives as larvae, sheltering underground or in old trees and rotting wood where they feed on decaying plant matter. This part of their life cycle can last for up to six years! By the time a stag beetle larva is ready to metamorphose into an adult beetle, it can be up to 4¼in (11cm) long. Once a larva has undergone transformation into a beetle in a chrysalis, its adult life will only last four months. They emerge in late May and are dead by August. Male stag beetles spend this time using their jaws to wrestle competitors for the chance to mate with females.

On balmy summer evenings, you can even see adult males flying through the air in search of mates. They fly upright with their wings stretched out behind them, making a buzzing sound.

STICK INSECTS
PHASMATODEA

Blessed with one of nature's most effective camouflage systems, stick insects blend in brilliantly with the bark, twigs and leaves of their forest home. There are more than 3,000 species of stick insects worldwide. The stick insect's scientific name, Phasmatodea, comes from the Greek word 'phasma', or ghost, and reveals its SAS-style survival skill of being able to mimic a host plant by swaying in the breeze, like leaves. Some exotic species even have fake buds and patches that look like lichen to give enemies the slip.

SMALLEST STICK INSECT:
Timema cristinae, North America, ¾–1¼in (2–3cm).

LONGEST STICK INSECT:
Phryganistria chinensis, China, up to 25in (64cm).

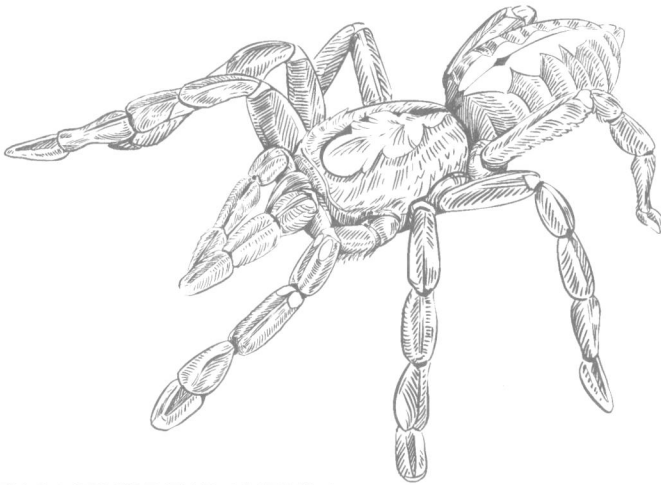

GOOTY SAPPHIRE TARANTULA
Poecilothera metallica

These critically endangered arachnids live in tall trees in the deciduous forest of central southern India – trees that are regularly chopped down for firewood and logging. Luckily, these tarantulas have an impressive jump and can also float down to the forest floor, leading to their other name: the peacock parachute tarantula.

MAMMALS

Around 68 per cent of mammal species make their home in forest and woodland habitats, from tiny shrews and bats to the apes that are our closest relatives. While some native species are under threat from introduced competitors, others are being reintroduced to their former habitats.

RED SQUIRREL
Sciurus vulgaris

Once common throughout Europe and Asia in coniferous, broadleaf and mixed woodland habitats, the red squirrel's population in the UK, Italy and Ireland was decimated by the introduction of the non-native grey squirrel in the late 19th century. The bigger greys outcompeted the smaller reds for food, and also spread the squirrelpox virus, which is fatal for red squirrels.

RED OR GREY? In theory, the orange-red fur ought to make the red squirrel easy to tell apart from its grey nemesis. In reality, the red squirrel's fur can vary from ginger to dark brown and often turns grey in the winter months. A way of differentiating the two species is to check the ears: reds often have large tufts above them, while greys never do. Also, greys are a lot bigger – around 19oz (540g) to a red's 10½oz (300g). A red's bushy tail is almost as long as its body.

Red squirrels mostly live on seeds and nuts, but occasionally make off with an egg or young bird. The best time to spot them is in autumn as they collect a stockpile of seeds and nuts to eat in winter when food supplies are scarce.

THE PURSUIT OF LOVE Male reds will chase females through the trees before they finally mate. Reds build a nest called a drey, high up in tree branches and assembled from twigs and moss, where their young are born. Within about 10 weeks of birth, infant squirrels are ready to forage for their own food.

ORANGUTAN
Pongo pygmaeus

Orangutans share 97 per cent of their DNA with humans and in the Malay language, orangutan means 'old man of the forest'. Physically, these orange-furred apes have long, powerful arms and hook-shaped hands to help them swing branch to branch through the rainforests of Sumatra and Borneo.

These awesome apes are huge in size and very intelligent. Adult males can grow to 4½ft (1.4m) in height and can weigh up to 286lb (130kg), normally twice the size of females. Because of their weight, male orangutans sometimes travel for miles on the forest floor, moving around on their hands and feet, as smaller tree branches aren't strong enough to support them! Male orangutans are solitary creatures and prefer to live alone. When moving through the rainforest, they howl to warn others to stay out of their way.

Orangutans sleep in trees, bending and weaving leafy branches together to make a comfortable bed to snooze on. In wet weather, they sometimes use large leaves to make a roof for their bed to keep themselves dry. They make a new nest every night, and conservationists count their nests to estimate their population size.

These clever creatures have opposable thumbs and big toes, and fish termites and ants out of tree holes with twigs.

EUROPEAN BADGER
Meles meles

The largest land predator in the UK, these stocky, stripy-faced woodland mammals can reach up to 35in (90cm) in length. They are native to Europe and West and Central Asia.

Badgers often eat slugs, earthworms and other invertebrates, but will also snuffle out fruit, such as apples, plums and elderberries, as well as small mammals including hedgehogs and rabbits. Eighty per cent of a badger's diet is earthworms.

They're found across the UK, but most of their population is located in southern England in woodland and open country.

SETTING THE SCENE Badgers have surprisingly sophisticated living arrangements. They live together in social groups of four to seven in a network of underground tunnels and burrows called a sett. These are usually built by the badgers in woodland. The main sett is where the badgers live and raise their young, while the series of smaller outlying setts are used as shelter if the badger needs to retreat to safety while out foraging. Badgers will routinely clean out their sett, removing old hay, grass and other materials they've used as bedding to prevent an infestation of fleas and lice.

BEAVER

CASTOR

There are two beaver species, the North American beaver (*Castor candenis*) and the Eurasian/European beaver (*C. fiber*), with the European beaver being slightly bigger at up to 31in (80cm) in length. Once widespread but having been hunted almost to extinction by the 20th century, these large – and charming – rodents fulfil an invaluable function. Living in streams and rivers next to woodland, they use their massive orange teeth to cut down trees and chop them into branches, which they pile up to create dams. By restricting streams' and rivers' water flow, they're able to produce ponds of deep water in which they construct lodges where they can live without being harassed by predators. These dams are enormously beneficial, creating new wetland habitat that can sustain many other species. They also help to prevent flooding.

It's commonly believed that beavers eat fish, but this is a misconception. In fact, they dive for aquatic plants and grasses – and they're able to hold their breath for up to 15 minutes while foraging underwater.

BIRDS

Around 60 per cent of bird species may be dependent on forests for their survival, as they feed on the plants, insects and bugs that thrive in them. Hearing their songs as we walk through the trees is one of the joys of a woodland hike.

WOODPECKER
PICINAE

There are about 180 species that belong to the subfamily of woodpecker birds called Picinae, which are found living among the trees in forests and woods across the globe, except for Australasia. They're famous for their drumming, the short sequence of tap-tap-tap sounds made by a woodpecker as it hammers its beak into tree bark. Primarily done to peck holes to access tree sap for feeding, drumming is also used to attract a mate and to mark their territories.

All that hammering would cause a woodpecker serious brain damage if its head wasn't specially adapted to protect this vital organ. The skull is made from a sponge-like bone that cushions the brain during rapid pecking. Their brains also fit very snugly inside their small skulls, which stops the grey matter from bouncing around. Thick muscles in their necks provide further shock absorption.

Woodpeckers help to keep trees healthy by devouring insects such as ants, caterpillars and borers.

CUCKOO
CUCULIDAE

There are about 60 species of cuckoo worldwide, living in forest and woodland.

Building a nest and looking after the eggs you lay takes a lot of work, so why not pass the job onto someone else without them realizing? This is the trick of some cuckoo species, who lay their eggs in the nests of other birds. It's called brood parasitism. These cuckoos choose the nests of birds that look similar to them, so the victims won't suspect they're being misled. Some cuckoos lay eggs that resemble those of other species to avoid detection.

When the baby cuckoo hatches, it pushes the other eggs out of the nest and is fed and looked after by its adoptive parents, who think they're raising their own chick. Since adult cuckoos have offloaded their parental duties onto other birds, they're free to migrate earlier in the season than other species. Altogether, it's a cruel but clever scheme!

MAKING MUSIC

Something you're almost guaranteed to hear a lot of in your local woodland is birdsong. It signals the breeding season, with males using birdsong to attract females and also mark and defend their territory. Birds generally use their syrinx — the avian version of a human's larynx or voicebox — to make the sounds used for communication. It sits at the base of their windpipe. The most complex syrinxes are found in songbirds, which can control the left and right halves of this vocal organ independently, effectively singing with two independent voices and thereby creating some remarkable sounds.

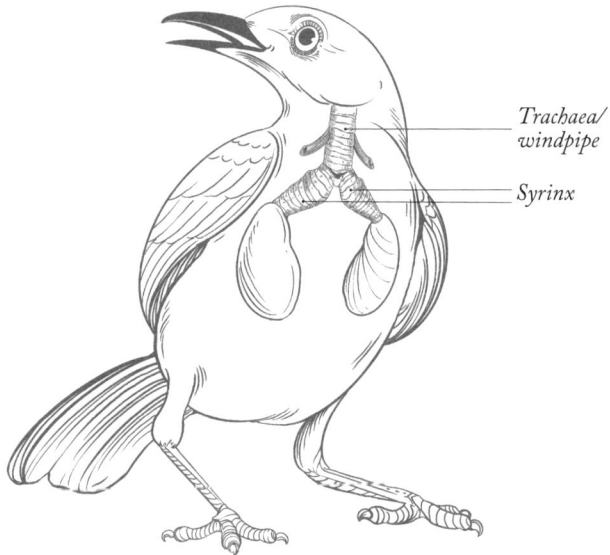

Trachaea/ windpipe

Syrinx

STARLINGS
Sturnus vulgaris

Starlings are amazing impressionists that copy the sounds they hear and work them into their lively songs. The philosopher Pliny claimed that in ancient Rome, starlings were taught to speak Latin and Greek! As far-fetched as it sounds, starlings can mimic human speech – but they don't understand it.

OWL
STRIGIFORMES

There are more than 200 species of owl worldwide (except Antarctica), many living in woodland habitats. Around 69 per cent of owls are nocturnal, hunting at night and sleeping during the day.

Owls are birds of prey and often eat small mammals for dinner, using their sharp talons to catch them. They have amazing vision and super-powered hearing that allows them to spot the quietest mouse from far away. Owls have asymmetrical ears (one is higher up than the other), so they can easily figure out where a sound is coming from and pinpoint their prey. Their flat faces act like a radar dish, directing sounds towards their ears. Owls have fringes on the tips of their flight feathers that make them virtually silent when they fly.

Most owls swallow their prey whole and then spit out a pellet made up of the bones, fur and feathers that they can't digest.

A REAL EYE-OPENER Owls don't have round eyeballs. They're more like elongated 'eye tubes', which help them to see in low light. While we can move our eyes from right to left, owls' eyes are fixed in one place, pointing ahead. And they're massive in proportion to their heads. Some species have eyes that weigh 5 per cent of their total body weight! Their vision is long-sighted, meaning they can't see things clearly close-up.

In ancient Greece, Athena, the goddess of wisdom and war, was often shown with an owl by her side.

REPTILES

From the deadly king cobra to the humble grass snake, many reptiles make their home in our forests and woodlands, although the proportion of reptiles living in these habitats is thought to be less than for other classes of animals.

KING COBRA
Ophiophagus hannah

Native to the forests of south and southeast Asia, the king cobra is known for its distinctive hood and hair-raising venom. An adult male king cobra is able to lift a third of its body length vertically off the ground in order to look an adult human in the eye. What's more, they can still move forward while doing this – scary stuff!

To deter predators and unwanted visitors from coming any closer, the king cobra will expand its ribs and muscles on either side of its head to create a hood that increases its size. Large males can make a hissing noise that sounds like the growl of a dog.

It's not so much the potency of their venom that makes a king cobra deadly, as the volume they can deliver: up to 1tsp (5ml) in a single bite. This massive injection of venom is powerful enough to kill a fully grown elephant!

Despite their fearsome reputation, king cobras only attack if they're cornered, provoked or protecting their eggs. They prefer to escape – and cause fewer than five human deaths a year.

They're the only snakes that build their own nests.

CHAMELEON
CHAMAELEONIDAE

Chameleons are arboreal animals, meaning that they live in trees and bushes. One-third of the world's 228 species live in the rainforests of Madagascar, while the rest are found in Spain, Portugal, Africa and Asia.

Chameleon skin is actually transparent. It's the cells underneath, filled with pigments, that change colour. Signals from their brains increase or decrease the size of the pigments, releasing different colours that blend together, a bit like when you mix paint.

A chameleon's tongue acts like a rubber band, snapping out and retracting to catch prey. The interplay between the tongue's sticky pad, its muscles and a small bone called the hyoid allows the tongue to accelerate from 0 to 62mph (100km/h) in a hundredth of a second. A poor insect chilling on a leaf doesn't stand a chance!

LONGEST CHAMELEON:

✺ 23½in (60cm) – the length of Parson's chameleon (*Calumma parsonii*)

SMALLEST CHAMELEON:

✺ 2⅓in (6cm) – the length of Nosy Hara Leaf chamelon (*Brookesia micra)*

Adder

EUROPEAN SNAKES

Grass snakes (*Natrix helvetica*) are the biggest of all the snake species in Britain. They are often found in woodland and are normally spotted close to water (they are superb swimmers). Grass snakes are a grey-green colour and recognisable by the yellow and black markings around their necks. When they're not busy slithering about, they can be found snacking on toads, frogs, fish and even small birds and mammals, such as mice.

Adders (*Vipera berus*) also hang out in European woodland. Grey or brown in colour, you'll know if you've come across one by the unusual zigzag pattern on their backs. While they may look stunning, adders can bite and are the only venomous snake in Britain (aside from all those weird and wonderful varieties kept in zoos or imported by private owners, of course!).

And don't forget the humble slow worm (*Anguis fragilis*). They're often mistaken for snakes, but they're actually legless lizards. Unlike snakes, they can blink, have a notched tongue and not much of a neck, with their head blending into their body. They dine on a diet of slugs, worms, snails and spiders, burrowing into the ground between October and March to hibernate.

If a slow worm encounters a predator, one of their weird defence mechanisms is defecating – the smell is so bad that it can put predators off!

KOMODO DRAGON
Varanus Komodoensis

The Komodo dragon uses its green skin to blend in with the bushes and tall grasses, waiting patiently for their prey. When something comes along, it uses its powerful, bowed legs to pounce, seizing the animal with their sharp claws and biting with their jagged teeth.

If its prey is bitten, but manages to get away, it won't last long: Komodo dragon saliva contains large amounts of harmful bacteria that infects the animal's wound, poisoning and killing them within a few days. All the Komodo dragon has to do is follow its injured prey using its fantastic sense of smell and wait for it to drop dead. Then it's dinner time.

* Up to 10ft (3m) long.
* Around 154lb (70kg) in weight.
* Eats up to 80 per cent of its body weight in one sitting.

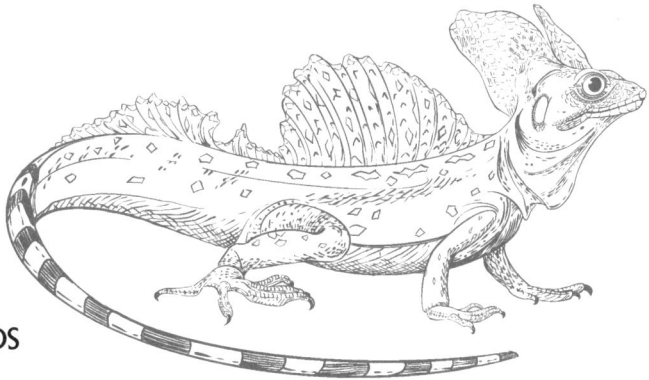

BASILISK LIZARDS
Basiliscus basiliscus

Native to the rainforests of Central and South America, these are also known as 'Jesus lizards' because of their amazing ability to sprint across water. If it senses danger, a basilisk can leap down from a tree and quickly build up enough momentum to run across water and escape from a predator. Basilisk lizards have large rear feet, with fringes of skin between their long toes that spread out on the water, providing a greater surface area and preventing the lizard from sinking. As they pedal their legs fast on the water, their feet push the water out of the way, allowing them to speed along like something out of a Looney Tunes cartoon.

CHAPTER THREE

WOODLAND CRAFTS

'Nature is the source of all true knowledge. She has her own logic, her own laws, she has no effect without cause nor invention without necessity.'
LEONARDO DA VINCI

'Adopt the pace of nature: her secret is patience.'
RALPH WALDO EMERSON

'Nature does not hurry yet everything is accomplished.'
LAO TZU

'Sadly, it's much easier to create a desert than a forest.'
JAMES LOVELOCK

M aybe you're the kind of person who loves to get stuck into a DIY project, or perhaps you run a mile if anyone so much as mentions fetching a hammer and nails. Either way, the next few pages contain woodland-themed practical projects to suit every sensibility.

With easy-to-follow instructions, plus a good set of tools and a little elbow grease, you can knock up a tasteful and unobtrusive shelter for your local birds. If you're after something a little more sedate, why not try out the tips for whittling and flower-pressing?

There's also a brief guide to the arts of coppicing and pollarding, explaining why these practices are so conducive to good woodland management.

Alternatively, if you're not in a creative mood right this minute, you can always skip ahead to the next chapter. The beauty is that these words will still be waiting for you, whenever you are ready.

MAKE A BIRD HOUSE

Birds make their own nests, so why do they need nest boxes? It's because there are fewer safe places in the wild for birds to build nests where they can feed and protect their young. By providing a dry and warm nest box, you are giving them a safe space for nurturing their chicks.

Different birds need different sized nest boxes and entrance holes. The ideal hole is just big enough for the bird to enter but too small for predators to get in. Find out about local birds that could nest inside your box, and what hole size they would need. Here are some examples:

BLUE TIT – 1in (2.5cm)
GREAT TIT, CHICKADEE, HOUSE WREN – 1¹⁄₁₀in (2.8cm)
HOUSE SPARROW – 1¼in (3.2cm)
TREE SWALLOW – 1⅓in (3.5cm)
EASTERN AND WESTERN BLUEBIRD – 1½in (3.8cm)

YOU WILL NEED:
* Length of softwood
* Hammer
* Handsaw or power saw
* Drill
* Drill bit
* Hole saw or flat wood drill bit
* ⅜in x 1⅛in (1 x 3cm) long screws or nails
* Piece of old rubber to make a waterproof hinge
* Pencil
* Ruler

6in
(15cm)

Side

8in
(20cm)

10in
(25cm)

10in
(25cm)

8in
(20cm)

Side

Base

4¾in
(12cm)

Back

18in
(46cm)

Front

8in
(20cm)

Roof

7in
(18cm)

Angle cut

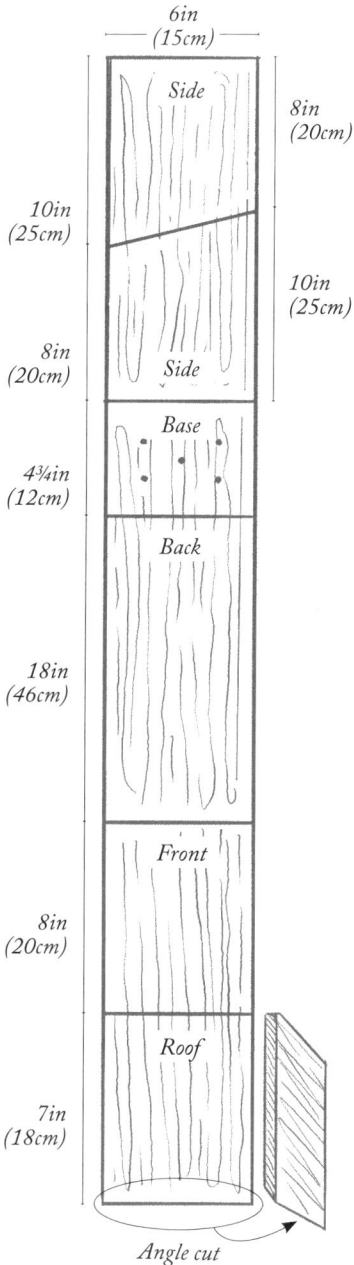

1 Measure and mark the pieces and then cut them out with a hand saw or power saw. Make sure you cut the pieces accurately to ensure that the box will fit together well. The plan shows you the size of the pieces you need. One of the short edges of the roof will need to be cut at an angle so that it will be flush with the board at the back.

2 Drill the entrance hole on the front piece. To do this you can either use a hole saw or flat wood drill bit. You also need to drill several small holes in the base to allow for drainage.

3 Using three evenly spaced nails or screws, attach the first side to the back. It should sit halfway up the back.

4 Next attach the base to the side using two nails or screws.

5 Now attach the other side to the back using three nails or screws in the back.

6 Attach the front to the two sides using three nails or screws each side.

7 The lid is the last piece of the bird box, and this is fitted to the back using a piece of rubber nailed to the roof and back to create a hinge. The rubber will keep the nest box waterproof.

TIP
Buy a piece of softwood that is straight and that has few knots. It should be at least ½in (1.5cm) thick.

8 You can now treat the box with linseed oil or wood preservative but be sure that it is non-toxic and safe for birds.

9 Mount the box on a tree or post with nails. It should be sheltered from day-long sunlight and wind, and between 6½–16ft (2–5m) above the ground (some species require their box to be higher). To avoid attracting predators, don't paint it in bright colours or place it too close to feeders.

CLEANING YOUR NEST BOX In the autumn, after birds have nested, you can remove and clean the nest box. Wash it out with boiling water to get rid of any parasites. Leave the lid up to let it dry.

MAKE A BESOM

Traditionally made from birch twigs with a willow or hazel handle, besoms were for centuries a commonly used broom, often associated with witchcraft. Used sideways with a swishing motion, these brooms are easy to make.

YOU WILL NEED:

* Twigs for the broom 2–3ft (60–90cm) long, ideally birch twigs
* A straight branch for the handle around 4ft (1.2m) long and roughly 1½in (3.8cm) in diameter
* Binding material – twine or wire
* Axe or pruning shears
* Whittling knife
* Nail or wooden peg
* Hammer

1 Sort your twigs by length.

2 Tie the longest twigs tightly together at the base. Arrange the shorter twigs around the outside of the central core and tie the whole head together, starting around 1–2in (2.5–5cm) from the base and wrapping the binding around tightly several times. Bind again a few inches further up until the twigs are firmly secured.

3 Whittle the narrowest end of your branch to a point, and push the pointed end into the middle of the broom head base. Then hammer a nail or wooden peg through the head and the handle to secure.

4 Trim off any stray ends of your besom with either an axe or pruning shears.

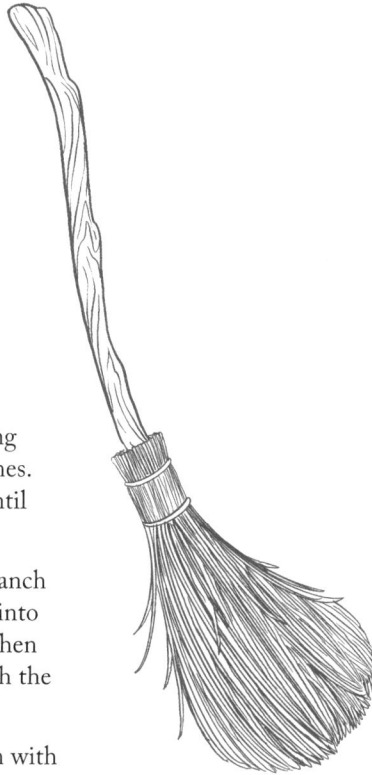

WHITTLING FOR BEGINNERS

One thing you're never far from in woodland is fallen twigs and branches, so why not get creative with this readily available material by learning the basics of whittling? As long as you follow some basic rules, it's a safe and rewarding activity that won't harm you or your precious local woodland.

GATHERING YOUR MATERIALS

Softer woods such as willow, sycamore and lime are much easier to carve. You'll need a smooth twig or branch without too many knots, and it must be young, fresh wood that hasn't dried out. Don't pull or cut healthy twigs or branches off a tree – instead, look for fresh ones on the ground.

Obviously, you'll need to bring your own whittling tools along with you as well. A vegetable peeler is excellent for stripping off bark, while a folding penknife is ideal for the fiddlier stages of your whittling. Some sandpaper can also come in handy for smoothing out rough edges.

Remember: carrying knives is against the law in many countries. In the UK, for example, carrying a lock knife is illegal unless you have a good reason to carry one. You're within your rights to carry one to the woods to use there, but if you stop off somewhere on the way, you would be committing an offence.

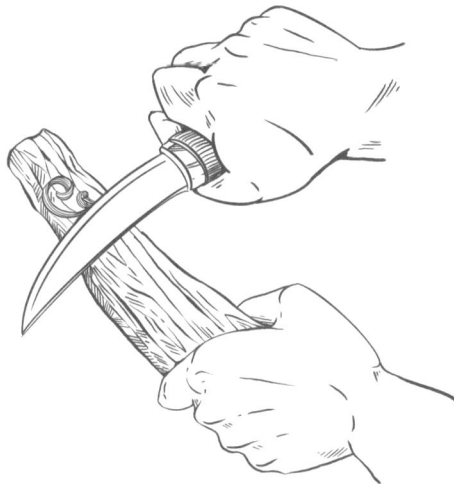

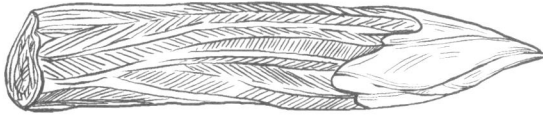

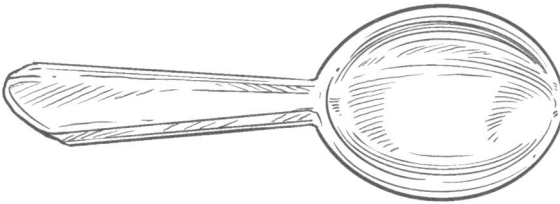

HOW TO WHITTLE SAFELY

You might want to practise using your peeler and knife on something softer and more pliable, such as a vegetable or soap bar, before upgrading to wood.

Always cut away from your body and the hand holding the wood, using a push stroke.

Don't hold the wood in your lap near the large blood vessels in your thighs, in case you slip and cut or stab yourself. Sit at a table when whittling or hold the wood past your knees or to the side.

Move slowly, whittling a tiny piece at a time. If you go quickly, you're more likely to slip or get your knife stuck.

Don't get too ambitious early on. You might want to start by whittling something simple like a toasting fork, but once you're a confident and experienced whittler there is no end to the possibilities.

PRESSING FLOWERS

Picking and pressing woodland flowers is a great way to learn about local plant species and to produce unique decorative craft projects, such as birthday cards or framed pictures. However, there's more to the process than you might think. Here's a brief guide to this rewarding and therapeutic woodland hobby.

Bear in mind that you must always have permission from the landowner before you start to pick and that plants in protected areas must not be picked at all. Check which rules apply in your country first.

BE ENVIRONMENTALLY AWARE

First things first, if you plan on picking wildflowers, then you need to make sure that you're doing so responsibly and staying within the law. Always practise moderation – don't collect so many samples that you're harming the plants. The 'one in 20' rule is a good way to approach it: if there are 20 plants, then it's fine to take a sample from one of them.

YOU WILL
NEED:

YOU WILL
NEED:

❋ Books

❋ Newspaper

❋ Card

❋ PVA glue

❋ A pen

1 Collect a flower sample from your woodland plant of choice. Try not to damage the plant or take too many samples from it.

2 Line an open book with newspaper. Place your flower on one page, laying it as flat as possible.

3 Now carefully and slowly close the book, weighing it down with other heavy books. It's important to have an even weight across the whole flower; if part of it is left exposed to the air, it will shrivel up. If you're pressing a heavier or thicker plant, you might want to add extra paper or card to the inside of the book to ensure that every bit of the sample is being pressed equally.

4 Place the pile in a warm, dry place. Check on it every day to see how things are proceeding. You want to dry the flower quickly, as this helps to preserve its colour. A space next to a radiator or central heating boiler is ideal.

5 Once the flower is completely dried out and preserved, carefully remove it. Depending on what type of plant you're pressing, this might take anywhere from a couple of days to a couple of weeks.

6 Now you can use the dried flowers to make an artwork, by mounting them on card using PVA glue.

COPPICING & POLLARDING

COPPICING

This ancient woodland management technique involves cutting a tree down to its base to create a 'stool' from which new shoots can grow. It can be used as a way of harvesting wood from a tree without irreparably damaging or killing it. Coppicing has several things going for it:

1 It allows for the gathering of wood that can be put to good, practical use as fuel or building material. Traditionally, hornbeam wood was used to make charcoal, chestnut for fencing and hazel for thatching.

2 The coppiced tree can regrow over a period of years, removing the need to constantly plant new trees. It's done on rotation, with small areas of woodland coppiced in turn each year. This routine allows the individual trees to live and produce wood for hundreds of years, creating a more sustainable model.

3 Biodiversity within the woodland can be increased, owing to the different ages of the trees – which attract different kinds of wildlife – and the varying levels of light now able to reach the woodland floor, which allows more plant species to flourish. Rare species such as dormice rely on the dense understory that can be established in coppiced woodland for shelter, food and so they can travel around without being exposed to predators.

In the first couple of years after a tree has been coppiced, the new shoots are vulnerable to being eaten by woodland creatures such as deer and rabbits. To counteract this, high deer-proof fencing or banks of earth are constructed around the 'stools'.

POLLARDING

Pollarding is a similar technique, but involves cutting back the canopy of a tree almost to the trunk to produce a dense mass of branches from which new growth can sprout. Again, it can be used to prolong the period of time over which a tree can supply wood.

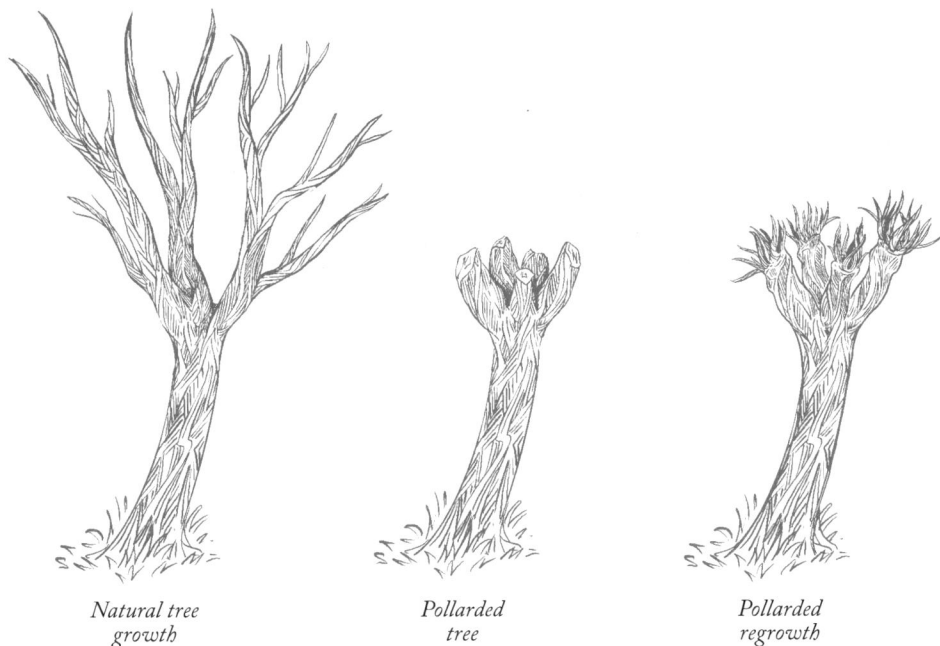

Natural tree growth

Pollarded tree

Pollarded regrowth

AN ANCIENT ART

There is some archaeological evidence that coppicing has been practised in Europe since the Neolithic period (4300–2000 BCE).

CHAPTER FOUR

FORAGING &
FOLK REMEDIES

'Leave the roads; take the trails.'
PYTHAGORAS

'Those who contemplate the beauty of the earth find reserves of strength that will endure as long as life lasts.'
RACHEL CARSON

'Choose only one master – nature.'
REMBRANDT

'The trees encountered on a country stroll
Reveal a lot about a country's soul.'
W. H. AUDEN

This chapter is for the braver reader, who wants to try his or her hand at the art of foraging. Forests are filled with naturally occurring delicacies, just waiting to be spotted, plucked and enjoyed – but be warned, you shouldn't go on the hunt without knowing the possible dangers.

It begins with a brief look at the basic rules of thumb worth following to make sure you and your fellow gastronomes don't end up making a trip to hospital – or the morgue. Remember: if in doubt, don't put it in your mouth! You'll also want to check the rules on foraging where you are to make sure you don't fall foul of trespass or foraging laws.

Included here is a rundown of some of the most delectable plants, fruits and fungi you might encounter on your forest rambles and how to enjoy them safely. And, for the really adventurous, there are some tips on enjoying some invertebrate snacks of snails and insects.

Last but not least, there are some traditional folk remedies that draw their ingredients from the forest, including a herbal poultice for treating minor injuries. There's also guidance on what to do in the unlikely event that your woodland wanderings leave you a little bit lost.

SAFE FORAGING

As long as you're careful and not reckless, foraging can be a safe, healthy and fun pastime, although obviously not without its risks. These are the main things to remember to save yourself and your friends or family from coming to harm:

* Correct identification is absolutely key. You need to be 100 per cent positive that something is edible before you collect and consume it, so consult two or more reliable sources. If there's even a shred of doubt in your mind, don't put it in your mouth!
* Keep your specimens separate. If it turns out that, on closer inspection, something you've foraged is dangerous, you don't want it to have contaminated your other, harmless acquisitions.

Deathcap mushrooms

Button mushrooms

🌸 Bear in mind the location where you found your specimens. Plants and fungi growing in polluted areas, such as near waste ground, may have drawn up toxic chemicals from the ground.

🌸 You also need to pay attention to potential hazards in the kind of terrain you're exploring. Beware of common obstacles like low-hanging branches, rabbit holes, deep puddles or slippery mud that might cause minor injuries like a twisted ankle or bumped head.

🌸 If you're foraging in an intertidal zone, then you MUST consult accurate tide tables in advance so that you leave yourself the time to get in and out of the area before the waters rush back in. Don't risk drowning for the sake of some seaweed!

🌸 Wear sensible clothing: this means clothing that won't tear or catch on thorny bushes. In wet weather, waterproof trousers and boots will help keep you dry and mean that you can kneel on the ground to collect specimens. In summer, don't be tempted to wade into long grasses wearing tiny shorts – this is a good way of getting a tick bite, which might result in a nasty case of Lyme Disease or Rocky Mountain Spotted Fever. A brimmed hat can also provide protection from the elements, be it rain or glaring sunlight.

FORAGING LAWS

Before you head out to pick your chosen delicacy, you should check the laws around foraging where you are. You need to know both what you are allowed to collect and where you are allowed to collect it from. You may be allowed to forage on public land – such as in parks or publicly owned forests – but not on private land, where you might also be subject to trespass laws. In some places taking leaves, berries and fungi may be allowed but uprooting or removing whole plants is likely not to be.

In the UK, there is a general right to forage the 'four Fs': flowers, foliage, fruit and fungi. The only caveats are that it must be growing wild, you must be collecting it for personal use, not profit and – depending on where you are – you may need the landowner's permission. In the US, you will need a permit to forage in National Parks, Forests and Monuments.

EDIBLE PLANTS

WORMWOOD
Artemisia absinthium

Native to north Africa and temperate regions of Eurasia, but also now found across Canada and the United States.

IDENTIFYING CHARACTERISTICS: A small shrub, with grey-green branching leaves on stiff stems and branches. It has ochre-coloured flowers with small petals and an unforgettably pungent (and often overpowering) smell. Often found in out of the way nooks and crannies, wormwood can thrive wherever there is good light and dryish soil.

USES: Known for its extremely bitter taste, wormwood is mostly only used nowadays to produce vermouth and absinthe.

Vermouth is made by steeping wormwood and other herbs for a few days, then distilling the resulting mixture and adding it to a dry white wine. Absinthe is concocted from a distilled spirit infusion of wormwood, anise and other herbs and spices. The famous yellowish-green colour comes from adding flavourless plant leaves, such as brooklime.

Wormwood's qualities can also be lent to making bitters for cocktails by infusing the plant with vodka and other ingredients – such as orange peel – for extra flavour.

WARNING!

Wormwood, or at least the component thujone in the plant, is slightly toxic and the percentage used in absinthe is limited by law in Europe. Make sure to consult a reliable source for information about safe amounts to use in your homemade recipes.

THE GREEN FAIRY Absinthe was first manufactured commercially by Henry-Louis Pernod in 1797. Often referred to in historical literature as 'the green fairy', it was long believed to have psychoactive effects on the drinker.

In the early years of the 20th century, absinthe came to be considered dangerous, with imbibers reportedly suffering convulsions and hallucinations. As a result, it was banned in many countries. However, modern research suggests that it is no more hazardous than any other spirit, and it is currently both legal in much of the world and enjoying a revival in popularity.

Absinthe was such a popular tipple of choice for writers and artists in Paris that five o'clock became known as the green hour. Illustrious literary enthusiasts of the liquor included:

* Oscar Wilde
* Charles Baudelaire
* Ernest Hemingway
* Émile Zola

In the Book of Revelation in the Bible, the star which falls to earth and turns a third of its waters bitter is called Wormwood.

WILD RASPBERRY
Rubus idaeus

This is most often found in forest clearings in woodland, or along woodland trails and hedgerows. The leaves of the plant are broad and oval-shaped with a sawtooth edge. At the right time of year – summer and early autumn – you should be able to find large, dense stands of the fruit, ready for picking. If the berries pull away easily from the core, then that means they're ripe.

USES: You probably don't need to be told that raspberries taste delicious. But did you know that the leaves of the raspberry bush can be dried and steeped in water to make a tea that reportedly has medicinal properties, including alleviating the symptoms of diarrhoea and menstrual cramp? Well, now you do!

WILD BLACKBERRY
Rubus fruticosus

Growing in European woodland, hedges and scrub, the fruit of the blackberry, or bramble, ripens in late July and lasts until late October – if you can find any that haven't already been stripped bare by foragers that late in the season!

It can be hard to tell raspberries and blackberries apart since the plants and fruits can look similar in colour and ripen around the same time. However, if it's black and the white core remains inside the fruit when you pluck it, then it's a blackberry.

USES: Jams, crumbles, wines, whisky and even vinegar.

BRAMBLE LORE

* UK folklore warns against picking blackberries after Old Michaelmas Day (10 October), as the Devil is meant to spit on them on this date.

* Blackberries have been discovered in the stomach of the remains of a Neolithic man, showing their enduring popularity as a woodland snack.

HOW TO MAKE BLACKBERRY WHISKY

YOU WILL NEED:
- A lot of blackberries
- Granulated sugar
- Whisky

1 Put your blackberries in a sealable jar and add a tenth of their volume in sugar.
2 Pour in some cheap supermarket whisky until you've covered the blackberries.
3 Leave the mix somewhere cool and dark for six months, giving the jar an occasional shake.
4 Now strain the liquid out of the blackberries and bottle your delicious blackberry whisky. If it's not quite sweet enough for your taste, add a little more sugar.

WOOD-SORREL
Oxalis acetosella

Identifiable by its distinctive trefoil leaves with three heart-shaped lobes. Wood-sorrel has white flowers, each with five petals and purple veins. At night, the leaves fold up, while during the day they open out flat.

USES: The leaves are high in vitamin C and can be eaten raw. The roots can be boiled and consumed, tasting a bit like a potato.

The presence of wood-sorrel is often used as an indicator of ancient woodlands in central and southern England.

* Kiowa Indians chewed on wood-sorrel to alleviate thirst.
* Cherokee Indians chewed on the plant to treat mouth sores.

HAZEL
Corylus avellana

Native to Europe and western Asia, the common hazel can be identified most easily by the yellow catkins that hang in clusters from mid-February. It also has round to oval and hairy leaves that turn yellow before falling in the autumn.

USES: The nuts, of course! It's best to pick them once they're plump but while they're still green, then let them ripen at home in a cool, dry place. They taste delicious.

IT'S A KIND OF MAGIC In folklore, the hazel has long been associated with magic. A hazel rod is said to ward off malign spirits and can also be used for water-divining – an ancient method of walking over an area while carrying a forked stick or rod and using it to locate groundwater below.

NETTLE
Urtica dioica

Stinging nettles aren't just a potential woodland hazard worldwide – they can also be (carefully!) collected for use in homemade drinks and meals. Like other green leafy vegetables they're a good source of many vitamins and minerals. Pick the young, tender leaves in spring, before they flower, as they have a better taste. Make nettle tea by adding one cup of nettles for every two cups of water and bringing it to the boil. Remove from the heat and leave to steep for 5 minutes, then strain the water into a cup, and enjoy. Or you can steam and wilt nettles like spinach, puree in smoothies or turn into pesto.

IS IT SAFE? Most adults should be able to consume stinging nettles without experiencing side effects, but if you're taking any medication, it's best to check with your doctor before trying them.

BETTER SAFE THAN SORRY

If you're not sure what plant you're dealing with and can't make an identification, it's best to steer clear. Toxic plants will often have one or more of these characteristics:

* An 'almondy' smell in the woody parts and leaves.
* Clusters of three leaves.
* Spines, thorns or fine hairs.

FORAGING FOR FUNGI

When looking for fungi, careful identification of every species you spot is utterly essential. While there are many delicious edible specimens out there, there are also a number of often similar-looking species that could cause serious harm or a horrible death! Here are some of the more common and lip-smacking varieties, some of which you might be able to forage for yourself.

TRUFFLES
TUBER

These ectomycorrhizal fungi grow among tree roots, so you're highly unlikely to be able to locate one unaided. Professional truffle-hunters use dogs to sniff them out before digging them up (originally pigs were used, but they kept eating them before the truffles could be retrieved!).

Prized as a delicacy, truffles have a musky fragrance and a rich, mouth-watering mushroomy flavour. They can fetch hundreds of pounds, but there are more affordable ways to try them out. You can buy truffle oil or black summer truffles in ordinary supermarkets, perfect for drizzling over a salad or shaving over a bowl of fresh pasta.

In 2007, a 3³⁄₁₀lb (1.49kg) white truffle was bought by billionaire Stanley Ho at auction for $330,000.

GIANT PUFFBALL
Calvatia gigantea

A favourite of many foragers, giant puffballs certainly earn their name, typically reaching sizes of 12–14in (30–35cm) in diameter and often found growing in rings of twenty or so individuals. They're most common in grassland, not woodland, but their popularity means they had to be included.

Only young puffballs can be eaten, so check that the flesh is firm and pure white in colour throughout. You can establish this by gently poking one. A good rule of thumb is only to harvest those that are smaller than a football.

To eat a giant puffball, you need to peel off the skin, slice the flesh into thin strips and break it into pieces. It's best to fry it evenly on both sides using butter and salt for taste.

5ft (1.5m) – the American record for a giant puffball's diameter.

OYSTER MUSHROOM
Pleurotus ostreatus

These grow in small groups or tiers on dead wood, most commonly beech, from July to October. Cream-coloured and oval or round in shape, their flesh is soft and has a fibrous texture. If you want to gather this species, it's best to head out early in the season and look for a toppled beech tree.

PENNY BUN
Boletus edulis

Another highly sought-after mushroom, penny buns are typically found in oak and beech woodland. They look a little like bread loaves or rolls, hence the name, with a brown and dimpled cap and a swollen, white or pale-brown stem. Appearing from July through October, depending on rainfall, they should be cooked slowly, yielding a potent umami flavour.

SCARLETINA BOLETE
Boletus luridiformis

Common in beech, oak, mixed and coniferous woodland, this mushroom might look alarming but is entirely edible and quite tasty. However, its physical similarity to some other dangerously toxic species means that it's not advisable for a novice forager to seek it out. If you want to track it down, find someone experienced. Its distinguishing features include a brown cap and a yellow stem with red dots, which instantly turns dark blue-black when sliced.

Don't eat it raw – it contains a toxin that can cause stomach problems, so always cook before eating.

CHANTERELLE
Cantharellus cibarius

These can be seen carpeting the floors of beech, oak, birch, spruce and pine forests during the months of August, September and October, where they grow in moss and leaf litter. They are golden yellow in colour with an upturned, trumpet-shaped cap – and reportedly carry a faint aroma of apricots. Simmering gently in a pan will bring out the full richness of their flavour.

HEDGEHOG MUSHROOM
Hydnum repandum

One of the easiest edible species to identify, hedgehog mushrooms have a cap that is irregular in shape, with a leathery texture, and an underside covered in the small, delicate spines that account for its name. Dozens will often be spotted growing in rings on the floor of pine, spruce, beech and oak woodland. Their flavour is mild and nutty when cooked. Eaten raw, they are too bitter to be enjoyed.

HEN OF THE WOODS
Grifola frondose

Almost always growing on oak trees – but also beech, hazel and sweet chestnut – this fungus is structured in tightly clustered, leaflike tiers that are grey on top and white underneath. You'll want to make sure you've collected a young specimen – as it ages, hen of the woods develops a bitter taste and can be invaded by maggots. Erm… bon appétit!

EDIBLE INVERTEBRATES

If you're feeling especially adventurous, you might want to try gathering some of these invertebrates for an unconventional feast. A word of advice: it's best not to spring these treats on your dinner guests unexpectedly!

SNAILS
HELICIDAE

If you're a frequenter of French restaurants, chances are you've already tried a snail or two. If you fancy harvesting and cooking your own, here's what you need to do. Be warned: it's a lengthy and rather disgusting process.

GATHERING Snails like to hang out on the underside of long-leafed plants. If you place a wooden board or plank on some bricks or rocks a few inches off the ground over some damp soil and leave it overnight, when you check in the morning, you'll likely find quite a few snails clinging to the bottom.

You need to store the snails somewhere cool and dark, such as in a bucket with a lid so they can't escape. It needs to have breathing holes – and you should spray a little misty water in the bucket every day to hydrate your future meal.

PREPARING You need to purify the snails before you eat them, so they're safe to consume (after all, you've no idea what these snails might've been eating!).

Start by feeding them greens and herbs for a couple of days. Then switch to cornmeal or oatmeal for another two days. Once their excrement is white, you'll know that their systems have been purged.

Now comes the slightly cruel bit: you need to starve the snails for a day or two to give them time to empty their bodies of excrement. Note: you'll need to clean the excrement and slime out of their bucket every day (after a few days of this, you might start to regret having embarked on this culinary adventure).

COOKING Finally, it's time to cook your little beauties.

1 Start by bringing a large pot of heavily salted water to a boil.

2 Add the snails and cook for 3 minutes.

3 Drain the snails and rinse them with cold water.

4 Remove each snail from its shell using tweezers or something similar.

5 Bring another pot to the boil. This one should contain 3 parts water to 1 part distilled vinegar.

6 Add the snails and cook them until their slime has been removed. This will take about 3 minutes.

7 Now you can replace the snails in their shells, which you'll need to have cleaned using a pot of boiling water containing two tablespoons of baking soda. Snails or escargots are traditionally served with garlic butter.

WILD ANTS
FORMICIDAE

Various types of ants are edible, including wood ants. Their distinctive large anthills, made of pine needles, are easy to spot in woodland. You can catch and eat them live – although you have to be careful to bite into the ant before it bites into you! Oddly, ants are said to have the flavour of citrus fruit.

GRASSHOPPERS
CAELIFERA

Grasshoppers (and also crickets) are in ready supply in woodland clearings and nearby heath or grassland. Grasshoppers are a highly popular snack across Asia and Africa – and rich in protein. They're often fried, mixed in with vegetables and salt. Apparently, they give off a beefy aroma when cooked.

FOLK REMEDIES

In the centuries before the advent of modern medicine, people relied on strange – and sometimes painful – traditional cures dispensed by physicians and wise women to cure their ailments. Many of these peculiar treatments involved trees and other substances gleaned from woodland, such as:

* Inhaling the smoke of a burning pine tree as a cure for the 'half-dead disease' – in other words, a stroke.

* Wearing shoes filled with tansy leaves to treat ague, a form of malaria that caused shivering fits.

* A cure for warts: pricking the wart with a pin, then sticking the pin in an ash tree and reciting the rhyme 'Ashen tree, ashen tree, Pray buy these warts from me' to transfer the affliction to the tree. Have you ever seen a warty tree? No? Well, it's probably a sign that this wasn't a very effective remedy.

* Toothache: similar to the warts cure, this involved hammering a nail into the tooth until it bled and then forcing the nail into a tree to transfer the pain away from the body. Poor trees!

PLANT CURES

ASH Ashes have long been regarded as healing trees. A magic ritual in Hampshire, UK, involved passing a child with weak or broken limbs through the split trunk of an ash tree. If the ritual worked, it was meant to knit their bones back together. Native Americans used white ash bark, leaves and seeds for a variety of medical purposes, ranging from a general tonic after childbirth to an aphrodisiac, an appetite stimulant and a cure for fevers.

ELDER The wood from an elder tree was said to chase off vermin and clear up warts. With its antiseptic and anti-inflammatory effects, elderflower was a popular folk remedy for centuries, used as a medicine to alleviate the symptoms of mild complaints such as colds or arthritis.

ROWAN The wood of the rowan tree was kept as a pocket charm against rheumatism. It was also used to stir milk, in the belief that its magical properties would prevent the liquid from curdling.

TREATING ACHES & PAINS

NETTLE STINGS

If you have been stung by a nettle, try not to touch the affected area for 10 minutes. This gives the chemicals from the nettle's hairs time to dry on your skin, which makes it easier to remove them. After 10 minutes, wash the affected area with a clean cloth soaked in soap and water. Then, if you have any strong tape or wax hair removal strips, use them to try to strip the nettle hairs off your skin. A cool compress applied to the area can provide additional relief.

HERBAL POULTICE

Used for thousands of years as a treatment for pains, aches, boils, abscesses, nettle stings and insect bites, a poultice is a hot compress – a folk remedy that can be assembled using medicinal herbs gathered from a woodland excursion or from the contents of your kitchen cupboards. Many plants have been shown to have anti-arthritic, anti-rheumatic and anti-inflammatory properties.

HEALING HERBS
turmeric
onion
ginger
garlic
dandelion
cat's claw
eucalyptus

KITCHEN ITEMS
Epsom salt
aloe vera
activated charcoal
baking soda
milk
bread
coconut oil

> **WARNING!**
>
> It's possible that a poultice applied directly to the skin could cause an allergic reaction. Start by applying a small amount to a patch of skin before you use the whole poultice on the affected area.

A SIMPLE POULTICE RECIPE

This recipe uses ingredients and materials that you've probably got in your home, but you can also collect herbs from the wild if you want.

YOU WILL NEED:
2 tsp coconut oil
¼ small onion, sliced
1 garlic clove, chopped
1 tsp turmeric powder
1oz (28g) freshly chopped or grated ginger
Cheesecloth or cotton bandage

1 First, add the coconut oil to a pan on a low heat. Add the rest of the ingredients and cook them until almost dry, but not burnt.

2 Remove the pan from the heat source and place the ingredients in a bowl to cool. Once they're warm, but not hot, you can move to the next step.

3 Spread out your cloth or bandage and ladle the ingredients into the middle.

4 Tie up the cloth into a bundle with the ingredients safely stored inside.

5 Now place the poultice on the affected area of your skin for 20 minutes.

If you're applying a poultice to an open wound, ensure you're using a clean cloth if it's a compress. If the wound appears to be seriously infected, you should seek medical attention instead of applying a poultice yourself.

If it's a heated poultice, ensure it is warm rather than hot before applying it to avoid scalding yourself.

WHAT TO DO IF YOU'RE LOST IN THE WOODS

You've finally decided to take that woodland hike and lose yourself for a couple of hours or an afternoon. There's just one problem… you've wandered too far, can't get your bearings and are now a little more lost than you'd bargained for.

If this is a situation you ever find yourself in, don't panic. Here are some tips for how to avoid getting lost, and how to find your way back to civilization.

USE YOUR SURROUNDINGS

It's always good to take an accurate map with you of the local area if you plan on trekking out into the deep woods. You can compare this to the landscape features you see around you to try to get a better sense of your position. Use natural landmarks, such as rivers or hills, to pinpoint your location as manmade landmarks, such as buildings or plantations, are more likely to have changed since the map was created.

If you're lost, try to backtrack to the last feature you remember that you can confidently place on the map. From there, you should be able to get back on track. And above all, don't panic! Calm down, take your time and move slowly.

PATHS ARE NOT ALWAYS YOUR FRIENDS

Never assume that any path you come across will take you where you want to go – or at least take you somewhere safe and civilized. Many paths in woodland are made by deer or sheep grazing on the plant life, so if you start blindly following a path you could find that it peters out in the middle of nowhere.

START SLOW

If you're new to hiking in woodland, then it's a good idea to start by doing short and easy trails. This way you can familiarize yourself with the navigational skills needed before embarking on something more ambitious – and potentially riskier. Try not to jump in at the deep end!

START EARLY

If you're doing a long hike, it's better to get going early in the morning so that you're finished before night descends. Walking deep in the woods at night is generally a bad idea: you're more likely to lose your way or sustain an injury if you can't really see where you're going.

TELL A FRIEND

Let a friend or family member know where you're going, so that if you don't return they'll be able to notify the police to search for you.

PACK ACCORDINGLY

Make sure to bring plenty of water and snacks along with you. If you do get lost and wind up spending more time in the woods that you'd anticipated, you need to be able to stay properly hydrated and keep your energy levels up.

CHAPTER FIVE

FORESTS IN FOLKLORE

'And this our life, exempt from public haunt,
Finds tongues in trees, books in the running brooks,
Sermons in stones, and good in everything.
I would not change it.'
WILLIAM SHAKESPEARE, *As You Like It*

'A tree is what it is,
Being what it is because it must be just that.
A tree is a tree,
So innocent and harmless, yet so savage,
and you can't make it out better than it is.'
WALT WHITMAN

'Storms make trees take deeper roots.'
DOLLY PARTON

Unsurprisingly, given their global presence, forests factor heavily in the world's mythologies and folktales. This final chapter takes a look at how forest landscapes have fertilized the imaginations of everyone from the ancient Greeks to the modern creators of popular entertainment.

Versions of a tree of life are common in many cultures, and individual trees are often imbued with mystical forces, such as the tree of knowledge of good and evil in Judaism and Christianity. There are other important trees too, including the World Tree or Yggdrasil of Norse mythology and the beautiful but formidable tree nymphs of ancient Greece.

Forests also have a hold on our imagination and folk history, such as in the complex allegories of Little Red Riding Hood or Hansel and Gretel. Shakespeare's forests are anarchic arboreal spaces, while in J. R. R. Tolkien's high fantasy masterpieces forests are immense and mysterious. Winnie the Pooh and his companions, on the other hand, inhabit a quainter, more benign woodland.

The chapter concludes with a consideration of how forests have impacted popular culture, from gateways into magical worlds to depictions of wooded spaces as something menacing and monstrous, with all kinds of horrifying creatures lurking within.

FOREST MYTHOLOGY

Forests feature heavily in the myths and traditional stories told by cultures around the globe.

TREE NYMPHS

Many cultures throughout history have believed that trees have spirits, with the wellbeing of the tree often inextricably linked to the wellbeing of its spirit – and vice versa. These entities commonly take a human or semi-human form.

Some of the most enduring are the dryads and hamadryads of ancient Greek mythology, which resembled beautiful young women.

Hamadryads, in particular, were said to be closely connected to – or even indivisible from – their trees, so if harm or destruction befell such a tree, they would also be injured or perish. For this reason, it was believed that mistreatment of a tree and its residing spirit would be met with a terrible punishment from the gods or the hamadryad.

Different kinds of dryads belonged to different types of trees:
DRYADS: oak trees
EPIMELIDES OR MELIADES: fruit trees
MELIAE: ash trees
DAPHNAIE: laurel trees

According to some ancient Greek sources, dryads and hamadryads are said to be the longest living of all mortal beings.

These tree nymphs have long been a source of inspiration for painters, writers and poets. In the 19th century, artists in Europe and America, including members of the Pre-Raphaelite Movement, used sensuous depictions of dryads as a means of exploring the beauty and eroticism of nature. *La dernière dryade* (The Last Dryad) by Gabriel Guay is a characteristic example.

Hamadryad

WORLD TREES

Almost every culture in the world has developed its own image of the cosmos and in many this has taken the form of a tree, with its canopy reaching to the heavens, its roots burrowing into the underworld and the human realm balanced somewhere in the middle.

A world tree of this kind reflects the metaphysical concept of the axis mundi: the axis around which the world and the universe revolves, connecting the spheres of the gods, living men and women and the dead.

YGGDRASIL Probably the most well-known world tree today is Yggdrasil from Norse mythology. For the Vikings, this utterly colossal evergreen ash tree held up the nine realms, including:

ASGARD, the glorious home of the Aesir gods, at the top of the tree. It was a giant fortress that reached so high it disappeared into the clouds – helpful for keeping out the frost giants. The most famous building there was Valhalla, where the gods ate banquets and drank beer with the souls of courageous warriors.

MIDGARD, the realm of the humans. The gods built an enormous fence around it to protect the mortals. It was surrounded by an ocean that was patrolled by a terrifying sea serpent.

Yggdrasil

Nidhogg

HEL, a dark and gloomy underworld where most people went when they died. It also included a place called Náströnd where all the dishonourable dead people went – thieves, murderers and pretty much anyone the gods disliked.

A handful of strange creatures were said to live on the tree:

NIDHOGG, a ferocious dragon serpent who also answered to the name of 'Malice Striker'. The *Prose Edda* describes Nidhogg as living under the third root of Yggdrasil. This fearsome beast, who gnaws at the roots of the tree, is said to sometimes slither into Hel.

RATATOSKR, or rat tooth, a red squirrel who scampered up and down the tree carrying spiteful messages between the other beings who lived there. Some say he was a little gossiper who liked to cause trouble.

RELIABLE NARRATOR? Much of what we know about Norse mythology comes down to us from the work of the 13th-century Icelandic historian, poet and chieftain Snorri Sturluson. In his *Prose Edda*, he collected and recounted these stories from the Norse tradition, although many experts think that he embellished or invented parts of these myths himself.

FORESTS & FAIRY TALES

Fairy tales exploring the fear and fascination exerted by the deep, dark woods are prevalent in cultures across the East and West.

One known to all English speakers is the tale of Little Red Riding Hood, which continues to be retold and reinterpreted in novels, movies, musicals, manga and video games. Although the story's origins are shrouded in mystery, a seminal version is French author Charles Perrault's classic story 'Le Petit Chaperon Rouge' from 1697, which the Brothers Grimm – German folklorists and academics – rewrote in the 19th century.

Variations on this basic tale are almost endless. The most common version has a little girl cloaked in red pursued by a hungry and villainous wolf through the forest, who eats her grandmother, disguises himself in her clothes and waits for the unsuspecting girl to arrive. Typically, it ends with a heroic woodsman coming to the rescue of the girl and her grandmother, cutting open the wolf's stomach with his axe to save them from being slowly digested.

But there are versions where the wolf's stomach is filled with rocks, causing him to collapse under their weight. And other iterations where the eponymous heroine uses her own wits to outsmart and defeat her slavering antagonist. There are also versions from Italy where the villain is an ogre, not a wolf, and similar tales from Japan, China and Korea where the wolf is replaced by a tiger.

WHAT DOES THE TALE SYMBOLIZE?

✳ A warning for young children of the hazards of the outside world?

✳ A metaphor for sexual predation of women by wolfish men?

✳ An illustration of female coming of age, sexual awakening and menstruation? Note the blood red colour with which Riding Hood is associated.

In her collection of short stories, *The Bloody Chamber*, and her screenplay for the film *The Company of Wolves*, the feminist and magical realist author Angela Carter repurposed this tale to look at modern questions about female agency, sexuality and the animal aspects of human nature.

HANSEL AND GRETEL

Another forest-based fairy tale further popularized by the Brothers Grimm is the dark story of the unfortunate siblings Hansel and Gretel. Abandoned by their adult guardians in the forest, they find their way to a house made from candy and gingerbread where a witch is waiting to cook and eat them.

The modern folklorist and academic Jack Zipes has shown that the tale originated in the Late Middle Ages, a period when the unwanted children of starving peasants were often abandoned in forests.

PAUL BUNYAN

An enduring American contribution to the canon of forest folklore are the stories of Paul Bunyan, a 7ft (2.13m) tall lumberjack of prodigious strength. Along with his trusty companion, Babe the Blue Ox, who was said to be as strong as nine horses and seven feet wide across his horns, Bunyan had many adventures in a series of tales passed down orally and collected in print in the late 19th and early 20th centuries. According to one, the Grand Canyon was first formed when Paul dragged his axe along the ground behind him!

WILLIAM SHAKESPEARE'S WOODS

In Shakespeare's lifetime, in the late 16th and early 17th centuries, England was an even more wooded country than it is today, but the ancient forest of Arden – which had surrounded Stratford-upon-Avon – was already in decline. Perhaps Shakespeare's sense of the loss of this precious forest inspired his focus on green spaces in his writing.

In his plays, Shakespeare reworks real forests into places where growth and transformation can take place. Often, they're explicitly magical realms where everyday order and logic are suspended.

A MIDSUMMER NIGHT'S DREAM

In this fantastical comedy, the two young lovers at its centre – Hermia and Lysander – escape into the forest near Athens, where a hidden society of fairies, ruled by the fairy king Oberon and fairy queen Titania, awaits.

A mischievous sprite called Puck plays tricks on the various human and fairy characters, using a magical formula to make them fall in love with the wrong partners. Chaos ensues, until Puck is forced to set things right by Oberon and order is restored, with a happy ending for Hermia and Lysander.

In this play, the forest is an anarchic space where unjust authority, like that of Hermia's father, is subverted and the force of erotic love runs riot.

AS YOU LIKE IT

O, how full of briars is this working-day world!
(Rosalind, Act 1 Scene 3)

Like *A Midsummer Night's Dream*, the heroine of this comedy, Rosalind,

Puck

{ 122 }

WOODS IN ENGLISH IDIOMS

* 'touch wood' – an expression of hope that one's good fortune will continue
* 'can't see the wood for the trees' – an observation that someone has lost sight of the bigger picture
* 'not out of the woods yet' – an acknowledegment that a tricky or dangerous situation is not yet over

escapes from the oppressive and unjust court of her uncle to the 'Forest of Arden' – a place no doubt inspired by the real forest of Shakespeare's youth. Here, Rosalind is able to choose a new identity for herself, changing her name and gender, taking on the role of the male 'Ganymede'.

There are no fairies here, but the forest is still a place where order is upended. Once again, the ensuing disorder gives way to an uplifting and happy conclusion.

MACBETH

Macbeth shall never vanquished be until
Great Birnam Wood to high Dunsinane Hill
Shall come against him.'
(Third Apparition, Act 4 Scene 1)

In this classic tragedy Macbeth is assured by the treacherous witches that his bloodthirsty reign will not end until Birnam Wood moves. Believing this to be an impossibility, Macbeth feels certain of his position. But his enemy, Macduff, has his soldiers camouflage themselves with the branches of Birnam Wood, thus giving the impression that the trees are converging on Macbeth's castle. With the prophecy unexpectedly fulfilled, Macbeth's fate is sealed.

OTHER SHAKESPEARE PLAYS FEATURING FORESTS:
* *The Merry Wives of Windsor*
* *Timon of Athens*
* *Henry IV Part One*

FORESTS IN MODERN CULTURE

From portals into a magical realm to landscapes stalked by sinister supernatural entities, forests and woodland play an important role in our culture, including movies, books and musicals.

TOLKIEN'S FANTASTICAL FORESTS

'I have always for some reason, I don't know why, been enormously attracted by trees. All my works are full of trees... I should have liked to make contact with a tree and find out what it feels about things.' J. R. R. Tolkien

High-fantasy writer and academic J. R. R. Tolkien's much-lauded novels *The Hobbit* and *The Lord of the Rings*, set in the fictional world of Middle-earth, feature many ancient forest landscapes populated with wise, majestic and sometimes dangerous creatures. Tolkien frequently juxtaposed the age and goodness of these realms with the destructive, industrialized power of his villains to convey the importance of preserving and protecting the natural world from rapacious consumerism.

MIRKWOOD: in *The Hobbit*, Bilbo Baggins and the Band of Dwarves enter this hazardous forest, where they're attacked by giant spiders and then taken prisoner by wood-elves.

LOTHLÓRIEN: a beautiful, idyllic forest in *The Lord of the Rings* ruled by elves and the headquarters of their resistance to the evil Sauron.

FOREST OF FANGORN: in *The Lord of the Rings*, the hobbits Merry and Pippin meet Treebeard, the leader of the Ents – a race of giant, treelike beings – in this ancient forest.

Influenced by legend, folklore and fantastical literature of the Old English and Middle English periods, Tolkien's work in turn has inspired many fantasy novels that include woodland or forest as one of their settings, such as the Forbidden Forest in J. K. Rowling's *Harry Potter* series, Immanent Grove in Ursula K. Le Guin's *Earthsea* and the forest of Skund in Terry Pratchett's *Discworld* series.

STRIKING A CHORD

Forests have also made their way into that most lucrative of modern theatrical genres: the stage musical. American composer and lyricist Stephen Sondheim's *Into The Woods* weaves together the plots of several famous fairy tales set partly or entirely in woodland, including Little Red Riding Hood and Rapunzel.

OTHER FORESTS IN LITERATURE

Gormenghast series: this blackly comic work of fantasy by British visual artist and writer Mervyn Peake is set in and around a vast, decaying castle with a sprawling forest where pivotal moments in the narrative unfold.

Winnie-the-Pooh series: Inspired by the ancient Ashdown Forest in East Sussex, England, author A. A. Milne created the Hundred Acre Wood where his beloved characters Pooh, Eeyore, Piglet and Tigger live.

Mythago Wood: This World Fantasy Best Novel award-winning work by Robert Holdstock takes place in and around the mysterious Ryhope Wood, a magical forest that is bigger on the inside than on the outside.

Lolly Willowes: English writer Sylvia Townsend Warner's feminist classic is the tale of a middle-aged, unmarried woman who finds new purpose practising witchcraft – and making a pact with the devil – in local woodland.

DON'T GO DOWN TO THE WOODS TODAY

With their poor visibility, eerie quiet and remoteness from civilization, forests are frequently an irresistible choice of setting for filmmakers. Forests have every horror imaginable, from unseen, malevolent witches (*The Blair Witch Project*), body-possessing demons (*The Evil Dead*), serial killers on the rampage (*The Last House on the Left*), werewolves (*Dog Soldiers*), monsters (*A Quiet Place*), the undead (*Pet Sematary*) and mysterious beings (*The Village*) to name just a few of the many reasons why, if you're in a horror or thriller flick, you should avoid entering the woods at all costs!

FURTHER READING

Ancient Woods, Trees and Forests: Ecology, History and Management
Alper H. Çolak, Simay Kirca and Ian D. Rotherham (Pelagic Publishing, 2023)

Enchanted Forests: The Poetic Construction of a World before Time Boria Sax
(Reaktion, 2023)

Forest Folklore, Mythology and Romance Alexander Porteous
(Hesperides Press, 2021)

How to Read a Tree: Clues and patterns from roots to leaves Tristan Gooley
(Hodder Press, 2023)

The Forager's Calendar: A Seasonal Guide to Nature's Wild Harvests John Wright
(Profile Books, 2020)

The Tree Almanac 2024: A Seasonal Guide to the Woodland World
Dr Gabriel Hemery (Robinson, 2023)

Treasury of Folklore: Woodlands and Forests Dee Dee Chainey and Willow
Winsham (Batsford, 2021)

Trees: A Comprehensive Guide Adam Adams (AAA Publishing, 2023)

Trees and Woodlands George Peterken (Bloomsbury Wildlife, 2023)

*Tree Wisdom: The definitive guidebook to the myth, folklore and
healing power of Trees* Jacqueline Memory Paterson (Thorsons, 2011)

Witch's Forest: Trees in magic, folklore and traditional remedies Sandra Lawrence
(Welbeck, 2023)

Woodlands: Trees, Flowers and Fungi Rebekah Trehern (Field Studies
Council, 2021)

USEFUL WEBSITES

UK

WOODS, TREES & FORESTS
forestresearch.gov.uk
woodlandtrust.org.uk

FOREST FAUNA
forestryengland.uk
woodlandtrust.org.uk

WOODLAND CRAFTS
craftcourses.com
woodlandskillscentre.uk
woodlandtrust.org.uk

**FOR INFORMATION ON
COLLECTING WILDFLOWERS:**
bsbi.org/wp-content/uploads/
dlm_uploads/Code-of-Conduct-v5-
final.pdf

FORAGING & FOLK REMEDIES
foraging.co.uk
historic-uk.com/CultureUK/Folk-
Remedies/
wildfooduk.com
woodlandtrust.org.uk

FORESTS IN FOLKLORE
treesforlife.org.uk/into-the-forest/
trees-plants-animals/trees/
woodlandtrust.org.uk/blog/2021/04/
tree-folklore/

USA

WOODS, TREES & FORESTS
americanforests.org
fs.usda.gov
parktrust.org
wilderness.org

FOREST FAUNA
americanforests.org
apps.fs.usda.gov/forest-atlas/
lives-introduction.html
wilderness.org

WOODLAND CRAFTS
woodcraft.com

**FOR INFORMATION ON
COLLECTING WILDFLOWERS:**
fs.usda.gov/wildflowers/ethics

FORAGING AND FOLK REMEDIES
eattheplanet.org
feralforaging.com
https://www.legendsofamerica.com/
na-remedy/wildedible.com

FORESTS IN FOLKLORE
native-languages.org/legends-forest.
htm
legendsofamerica.com/az-
petrifiedcurse/

First published 2024 by
Guild of Master Craftsman Publications Ltd,
Castle Place, 166 High Street, Lewes, East Sussex BN7 1XU, UK

ISBN 978-1-78494-699-9

PUBLISHER Jonathan Bailey
PRODUCTION Jim Bulley
EDITOR Alexis Harvey
DESIGN MANAGER Robin Shields
DESIGNER Michael Whitehead
ILLUSTRATOR Alejandra Penaloza

Colour origination by GMC Reprographics
Printed and bound in China

To order a book, contact:
GMC Publications Ltd
Castle Place, 166 High Street, Lewes,
East Sussex, BN7 1XU
United Kingdom
Tel: +44 (0)1273 488005
www.gmcbooks.com

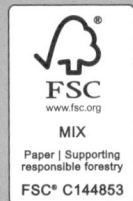

FSC
www.fsc.org
MIX
Paper | Supporting
responsible forestry
FSC® C144853